TIME

GREAT DISCOVERIES

Editor Kelly Knauer
Art Director Ellen Fanning
Picture Editor Patricia Cadley
Research Director Matthew McCann Fenton
Contributing Writer Matthew McCann Fenton
Copy Editor Bruce Christopher Carr
Production Director John Calvano
TIME Special Projects Editor Barrett Seaman

TIME INC.
HOME ENTERTAINMENT

President
Rob Gursha

Vice President, Branded Businesses
David Arfine

Executive Director, Marketing Services
Carol Pittard

Director, Retail & Special Sales
Tom Mifsud

Director of Finance
Tricia Griffin

Marketing Director
Kenneth Maehlum

Assistant Director
Ann Marie Ross

Editorial Operations Manager
John Calvano

Associate Product Managers
Jennifer Dowell, Meredith Shelley

Assistant Product Manager
Michelle Kuhr

Special thanks to: Victoria Alfonso,
Suzanne DeBenedetto, Robert Dente,
Gina Di Meglio, Peter Harper, Roberta Harris,
Natalie McCrea, Jessica McGrath,
Jonathan Polsky, Emily Rabin,
Mary Jane Rigoroso, Steven Sandonato,
Cristina Scalet, Tara Sheehan, Bozena
Szwagulinski, Cornelis Verwaal,
Marina Weinstein, Niki Whelan

Acknowledgments

Introduction by Diane Ackerman ©1997.
Reprinted from a TIME Europe special issue.
Used by permission of the author and the
William Morris Agency.

The Stephen Jay Gould quote on p. 3 is drawn
from a 1991 interview with TIME.

The Carl Sagan quote on p. 57 is excerpted
from *Pale Blue Dot* (Random House, 1994;
softbound edition, Ballantine Books, 1997).
Used by permission.

The Jacques Cousteau quote on p. 111 is from
a 1960 TIME cover story.

The Roger Payne quote on p. 141 is drawn from
an interview conducted in 2001 for this book.

First Edition

ISBN 1-929049-33-1
Library of Congress Number: 2001091489

TIME Books is a trademark of Time Inc.

We welcome your comments and suggestions about TIME
Books. Please write to us at: TIME Books • Attention: Book
Editors • PO Box 11016 • Des Moines, IA 50336-1016

If you would like to order any of our hard-cover Collector
Edition books, please call us at 1-800-327-6388 (Monday
through Friday, 7 a.m.–8 p.m., or Saturday, 7 a.m.–6 p.m.
Central time).

Please visit our website at www.TimeBookstore.com

Printed in the United States of America

Cover picture credits:

Comet Hale-Bopp (softbound): John Chumack—
Galactic Images/Photo Researchers
Mt Etna exploding (hardbound): Dr. Juerg Alean—
SPL/Photo Researchers
Machu Picchu (hardbound back cover): Brian Vikander—
Corbis

Inset cover photos:

Moonwalk: NASA; Whale: Iain Kerr—Ocean Alliance;
Chinese warriors: O. Louis Mazzatenta—NGS; Machu
Picchu: Brian Vikander—Corbis; Gorillas: Konrad Wothe—
Minden Pictures; King Tut mask: Luis Vollota—Bruce
Coleman; Saturn: NASA; Pyramid: Robert Frerck—
Woodfin Camp

TIME

GREAT DISCOVERIES

An Amazing Journey Through Space and Time

▰▰ contents

136

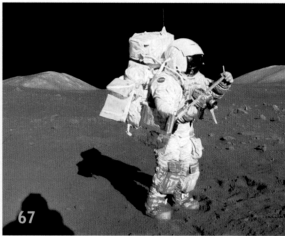

67

26

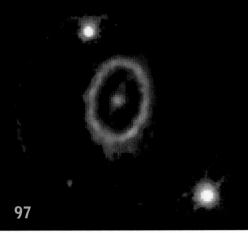

97

112

Discoveries on Your Doorstep By Diane Ackerman

WHEN I READ OF THE RECENTLY DISCOVERED *Symbion pandora*, a radically new lifeform that's pinpoint small, trisexual (it will try anything) and lives on the lips of lobsters, my first thought was: Do lobsters have lips? But that was quickly followed by a renewed sense of wonder at the quirky fantasia of life on Earth. With a mouth like a hairy wheel and other anatomical oddities, S. *pandora* is so outlandish that a special phylum was created for it—*cycliophora*, of which pandora is the sole member.

I must admit, I get a devilish delight when the miraculous appears right under my nose. After all, the marvelous is a weed species: one can glimpse it on one's doorstep. People often ask me where they might go to find adventure. But adventure is not something you must travel to find; it's something you take with you. The astonishing can turn up in the leaf clutter or even at a neighborhood restaurant, in a dingy tank, on the lips of lobsters.

We forget that the world is always more and stranger than we guess. Or can guess. Instead, we search for simple answers, simple laws of nature, in a sleight of mind that makes us uniquely human. Just as we're addicted to rules, home truths and slogans, we're addicted to certain ways of explaining things. There's bound to be a simple answer to everything, we insist. Maybe not. Maybe complexity frightens us. Maybe we fear becoming as plural as all we survey. Maybe we still tacitly believe that the universe was created for our pleasure, that we pint-sized demigods are its sole audience and goal. Then something like *pandora* turns up, a creature that breaks all the rules and gives biologists a jolt.

Because we have swarmed across the world with our curious and agile minds, we sometimes think that nature has been fully explored. But that's far from true. Plants and animals are becoming extinct at an appalling rate, and many of them are vanishing mysteries. The riches of the natural world are slipping through our fingers before we can even call them by name. Hanging on by a suction cup, and reaching around to vacuum up fallen morsels from a dining lobster's lips, *pandora* reminds us that we share our planet with unseen hordes, and it hints at the uniqueness of our own complex niche in the natural world. We know the tropics contain a rich pharmacopoeia, but there are many organisms still to be found in our own backyards.

Variety is the pledge that matter makes to living things. Think of a niche and life will fill it, think of a shape and life will explore it, think of a drama and life will extravagantly stage it. I personally find pampas grass an unlikely predica-

ment for matter to get itself into, but no stranger than we human beings, the lonely bipeds with the giant dreams.

The word discovery literally means: uncovering something that's hidden from view. But what really happens is a change in the viewer. The familiar offers a comfort few can resist, and fewer still want to disturb. But as relatively recent inventions such as the telescope and microscope have taught us, the unknown has many layers. Every truth has geological strata.

The moment a newborn opens its eyes, discovery begins. I learned this with a laugh one morning after delivering a calf. When it lifted up its fluffy head and looked at me, its eyes held the absolute bewilderment of the newly born. A moment before it had enjoyed the even-black nowhere of the womb, and suddenly its world was full of color, movement and noise. I've never seen anything so shocked to be alive.

Discoverers keep some of that initial sense of surprise lifelong, and yearn to behold even more marvels. Trapped in the palatial rut of our senses, we invent mechanical extensions for them, and with each new attachment more of the universe becomes available. Some of the richest moments in people's lives have come from playing with a mental box full of numbers or ideas, rotating it, shaking it, while the hours slip by, until at last the box begins to rattle and a revelation spills out.

And then there are those awkward psychological mysteries. I suspect human nature will always be like mercury, a puzzle near impossible to grasp. No matter how much of the physical universe we fathom, what makes us so quintessentially human eludes us, because it's impossible for a system to observe itself with much objectivity. Of course, that makes studying human nature all the more sporting.

I rarely dwell on this when I go out biking through the countryside each day. I don't worry about the mites that live among my eyelashes either. I have other fish to fry: the local land trust's campaign for acreage, the plight of endangered animals and landscapes, not to mention all the normal mayhems of the heart. But I get a crazy smile when I think of *pandora.* I like knowing the world will never be small enough to exhaust in one's lifetime. No matter how hard we look or where we look, even under our own or a lobster's nose, surprise awaits us. There will always be plenty of nature's secrets waiting to be told. This is one of those tidy, simple-sounding truths I mentioned, the sort of thing humans crave. And I believe it because I got it straight from a lobster's lips.

Diane Ackerman is a poet, essayist and naturalist who has written 15 books about nature, human and otherwise.

Discoveries in
Time

" There are about half a dozen scientific subjects that are immensely intriguing to people because they deal with fundamental issues that disturb us and cause us to wonder. Evolution is one of those subjects. It attempts, insofar as science can, to answer the questions of what our life means, and why we are here, and where we came from, and who we are related to, and what has happened through time, and what has been the history of this planet. These are questions that all thinking people have to ponder. "

—Stephen Jay Gould

COSMOS STUDIOS-MPH ENTERTAINMENT

On the Trail of A Lost Dinosaur

Move over, Indiana Jones. A quest that began with a game of "what if" over a round of beers—and involved searching for rare dinosaurs whose fossils were destroyed when Allied bombs smashed German targets in 1944—reached sweet fruition in 1999. As the journal *Science* reported in May 2001, a team of young researchers not only found examples of the lost specimens—they came away with the bones of a new species of dinosaur, believed to be the second-largest "terrible lizard" ever to walk the earth.

The location was Egypt's Bahariya Oasis, some 200 miles southwest of Cairo—the site of another important discovery of recent years, an ancient burial ground of pharaonic and Roman-era mummies (*see p. 34*).

The dinosaur hunters were led by Joshua P. Smith, a Ph.D. candidate at the University of Pennsylvania; they were following in the footsteps of a legendary dino sleuth, Ernst Stromer von Reichenbach, a University of Munich geologist and paleontologist, who had explored Bahariya before World War I and found four new dinosaur species. But Reichenbach's collection was bombed to dust during World War II. As Smith tells the story, he and his colleagues first concocted the notion of finding "Stromer's lost dinosaurs" when compiling a beer-fueled wish list of great achievements still to be accomplished in paleontology.

Smith and his team got their chance when they were given a few days leave from a larger expedition to Bahariya to conduct their own research. They began digging—and scored a find on their first day of searching. Returning the next year, they uncovered an enormous humerus, the upper arm bone of a dinosaur. Six feet in length, it was 14% longer than that of the next largest dinosaur of that geologic period, the Cretaceous. The new species was named *Paralititan stromeri*: the first term means "tidal giant"; *stromeri* honors the paleontologist whose life's work was redeemed decades after his death.

Photograph by Joshua B. Smith—U. of Pennsylvania

DEM BONES Graduate student Matthew Lamanna of the University of Pennsylvania excavates the left humerus of *Paralititan*

ON A DIG IN DINOSAUR HEAVEN

The discovery of *Paralititan stromeri* by a team led by Josh Smith, above, revived interest in the work of Ernst Stromer von Reichenbach, below. It also brought attention to the Bahariya region of Egypt, which scientists believe was a coastal region millions of years ago, as a heavily populated "dinosaur heaven."

The leaf-eating *Paralititan stromeri* was some 80 to 100 ft. long and weighed as much as 70 tons. It appears to be second in size among dinosaurs only to a South American titanosaurid, *Argentinosaurus huinculensis,* the most massive terrestrial animal. The large tooth of another dinosaur, possibly a carnivore that had scavenged the carcass, was found near the *Paralititan* remains. The fossils were dated at about 94 million years old.

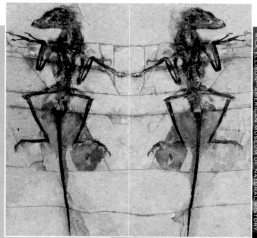

MICK ELLISON—AMERICAN MUSEUM OF NATURAL HISTORY

WHY THE FEATHERS?

The discovery of a well-preserved fossil of a feather-covered dinosaur in China (above, a sketch) helped cement the dinosaur-bird connection, but it has also raised the question of why dinosaurs needed feathers in the first place. Biologists once assumed that feathers evolved for flight. But as paleontologist Kevin Padian of the University of California, Berkeley, says, "The feathers on these dinosaurs aren't flight-worthy, and the animals couldn't fly. They're too big, and they don't have wings."

So what was the original purpose of feathers? They might have been useful for keeping dinosaurs dry, distracting predators or attracting mates. But many biologists suspect their role was to keep dinos warm. Dinosaurs' bone structure shows that, unlike modern reptiles, they grew as fast as birds and mammals —which dovetails with increasing evidence that dinosaurs were, in fact, warm-blooded.

The Thing With Feathers

The once radical notion that birds descended from dinosaurs—or may even *be* dinosaurs, the only living branch of the family that ruled the earth eons ago—has got stronger and stronger since paleontologists first started taking it seriously in the 1970s. Remarkable similarities in bone structure between dinos and birds were the first clue. Then came evidence, thanks to a series of astonishing discoveries in China's Liaoning province in the late 1990s, that some dinosaurs may have borne feathers. But a few scientists still argued that the link was weak; the bone similarities could be a coincidence, they said. And maybe those primitive structures visible in some fossils were feathers—but maybe not. You had to use your imagination to see them.

Not anymore. The find of a spectacularly preserved fossil of a juvenile dromaeosaur, announced in the science journal *Nature* in April 2001, by a team of paleontologists from the Chinese Academy of Geological Sciences and New York City's American Museum of Natural History, may be the long-sought link. "It has things that are undeniably feathers," exulted Richard Prum of the University of Kansas Natural History Museum, an expert on the evolution of feathers. "But it is clearly a small, vicious theropod similar to the velociraptors that chased the kids around the kitchen in *Jurassic Park*."

In fact, this small relative of *Tyrannosaurus rex*, dating from 124 million to 147 million years ago, has no fewer than three different types of feathers. The head sports a thick, fuzzy mat of short, hollow fibers, while the shoulders and torso have plumelike "sprays" of extremely thin fibers up to 2 in. long. The backs of its arms and legs are draped in multiple filaments arranged in a classic herringbone pattern around a central stem. Even the theropod's tail is covered with feathers, with a fan, or tuft at the end. *T. Rex*, meet Tweety Bird.

Illustration by Michael Skrepnick

DINO DUCK? He looms large in the illustration at left, but this *Caudipteryx* theropod— a cousin to the new find—was about the size of a duck. Scientists now think baby tyrannosaurs also looked much like this

British anatomist Richard Owen coined the term dinosaur ("terrible lizard") in 1841 to characterize gigantic fossilized bones and footprints found decades earlier and ascribed to dragons, extinct lizards, even giant ravens. The quest to unearth the remains of these intriguing prehistoric beasts is one of the great ongoing adventures in science

Edwin Colbert

American Museum of Natural History curator Edwin Colbert, above, found a massive trove of *Coelophysis* fossils in 1947 at Ghost Ranch, N.M. The swift predator, about 8 ft. long, had a long, slender neck, tail and hind legs; a birdlike posture; and a long, narrow head equipped with many sharp teeth.

Douglas Lawson

While still a graduate student, Lawson uncovered the humerus (upper arm bone) of a giant flying dinosaur that he dubbed *Quetzalcoatlus northropi,* at Big Bend National Park in 1971. With an incredible wingspan of up to 40 ft., this long tall Texan is the largest flying reptile, or pterosaur, found to date.

SHELLY KATZ—TIMEPIX

Jack Horner

Curator of paleontology at the Museum of the Rockies in Bozeman, Mont., Jack Horner found the first dinosaur eggs in North America and the first dino embryos ever unearthed. In 1982—a year after the picture at right was taken—Horner found his richest prize, a bone bed containing more than 10,000 duckbill dinosaur fossils.

JOE MCNALLY

BROOKS WALKER

Rodolfo Coria

Oh, baby! After years of searching, it's no surprise that dinosaur hunters feel paternal upon finding a millions-year-old fossil. At left is Argentine paleontologist Rodolfo Coria in 1988; one year later, he would find fossils of an enormous new species, *Argentinosaurus huinculensis,* in Patagonia. Weighing in at an estimated 100 tons, *Argentinosaurus* is the largest dinosaur discovered to date.

Sue Hendrickson

No, field paleontologist Sue Hendrickson didn't discover *Tyrannosaurus rex*, but in 1990 the amber specialist found the best-preserved fossil ever unearthed of the predator we love to hate, on the Cheyenne River Sioux Indian Reservation in the Badlands of South Dakota. At right, Hendrickson shares the spotlight in 2000 with the reconstructed model of her approximately 67 million-year-old find, installed at Chicago's Field Museum. The fossil of the "king of tyrants" is known to one and all—naturally—as "Sue."

JOHN ZICH—AFP-CORBIS

NATIONAL MUSEUM OF KENYA—NATIONAL GEOGRAPHIC SOCIETY

▬ profile

First Family of First Families

Sifting through the soil of ancient African savannas, the Leakey Family dug deep into the past, finding secrets in skulls and measuring eons in jawbones

AYBE IT'S PURE LUCK. MAYBE IT'S 90% perspiration and 10% inspiration. Maybe it's natural selection. Whatever the origin of the species of great fossil hunters known as the Leakeys, there is little doubt about the result: they found the keys to our past. The primordial Leakeys, Louis and Mary, conducted excavations at Tanzania's Olduvai Gorge that transformed paleoanthropology—the study of human origins—from a superficial examination of stones and bones into a rigorous discipline aimed at pinpointing the crucial steps on the evolutionary pathway to modern humans.

But their son Richard, heir to the kingdom, pulled a Prince Hal—he shunned his parents' world, even though he had found his first fossil, the intact jaw of an extinct giant pig, when he was only six. Instead, he ran a successful safari company and learned to fly. "I just wanted to get out—to go away and look at animals," he once said. "But it passed. The time came."

His parents had set the standard high. Louis, born in 1903 in Kenya to missionaries, was attending Cambridge when he joined his first expedition. At Lake Victoria in 1948 he discovered a skull of *Proconsul africanus*, an apelike ancestor of modern primates that lived 25 million to 40 million years ago. He and his wife Mary took their search for fossils to east Africa's Olduvai Gorge, in what is now northern Tanzania. There, in 1959, Mary discovered jaw fragments of a 1.75 million-year-old hominid, which they named *Zinjanthropus*. This apelike prehuman that first evolved some 4 million years ago is now called *Australopithecus boisei*. In 1960 the Leakeys and their eldest son Jonathan unearthed fragments of another hominid, *Homo habilis* ("Handy man"), which had a larger brain than the australop-

ithecines. They believed *Homo habilis* was the first tool-using hominid.

Richard broke free of his parent's shadow in 1968 when, without warning his father, he asked the National Geographic Society for a grant to explore some promising sediments on the eastern shore of Kenya's Lake Turkana. Like his parents' site at Olduvai, Turkana is part of the Rift Valley system, a giant fissure that runs north to south in eastern Africa for about 6,000 miles. Richard was blessed with "Leakey's luck"; his first hominid fossil, a jaw from *A. boisei*, turned up after just three weeks. Over the next 20 years—before Richard gave up hunting fossils in 1989 to run Kenya's National Wildlife Service—the fossils of early man his team found in the region eclipsed his parents' work. Among them:

● E.R.-1470, a 1.8 million-year-old skull that is one of the best-preserved specimens ever found of *Homo habilis*.

● "Turkana boy," the nearly intact skeleton of a strapping youth who lived at least 1.6 million years ago. He is the oldest and most complete fossil to date of a close relative of *Homo erectus*, the first human ancestor to establish campsites and master the use of fire.

● The "Black Skull," a 2.5 million-year-old cranium that is now thought to resemble the common ancestor of certain *A. boisei*.

Unhappily, Louis Leakey resented the acclaim that greeted Richard's early work; their intense conflict threatened to split the family. The feud ended in 1972 when the elder Leakey flew to visit his son's camp; by the harsh light of a gas lamp they examined fossils late into the night. Inspired, Louis predicted that Richard would find evidence of three hominid species at Turkana. A few weeks later, Louis died, unaware that events would prove him right.

2001: THE LEAKEYS' LATEST FIND

Move over, Louis, Mary and Richard Leakey: here come Maeve and Louise, Richard's wife and daughter. In April 2001, they announced that at Kenya's Lake Turkana in 1999 they had found the partial remains of a humanlike skull. The find reignited a great debate: Did modern man evolve in direct steps from a common apelike ancestor? Or did the human family tree sprout several limbs, some of which died out?

Most paleontologists believe that there was just one hominid line, starting with a small, upright-walking species, *Australopithecus afarensis*, most famously represented by "Lucy," the skeleton found in Ethiopia in 1974. The new skull challenges that view, pushing the presence of coexisting species back another million years—right into Lucy's time. Yet it is so different from Lucy that the Leakey team assigned their fossil, dubbed *Kenyanthropus platyops*, or "Flat-faced man of Kenya," to a new genus, or grouping of species. Is old flat-face a direct link in the human line? Or is he part of a branch leading to similar species? To find out—as the Leakeys might say—dig we must.

Top: Photograph of Richard Leakey, 1972—Popperfoto-Archive-Liaison
Left: Photograph of Louis Leakey, 1965 by Ian Berry—Magnum Photos
Right: Photograph of Mary Leakey, 1978 by Bob Campbell—Corbis Sygma

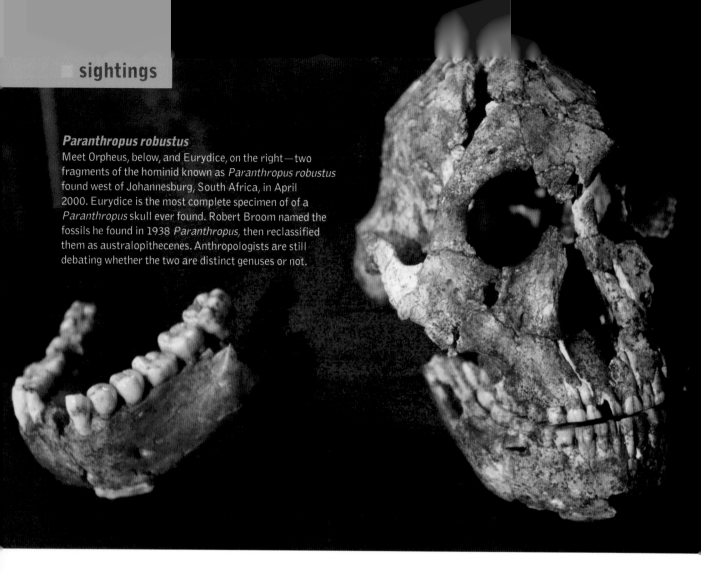

Paranthropus robustus

Meet Orpheus, below, and Eurydice, on the right—two fragments of the hominid known as *Paranthropus robustus* found west of Johannesburg, South Africa, in April 2000. Eurydice is the most complete specimen of of a *Paranthropus* skull ever found. Robert Broom named the fossils he found in 1938 *Paranthropus,* then reclassified them as australopithecenes. Anthropologists are still debating whether the two are distinct genuses or not.

Talking Heads

Legbone connected to the kneebone, kneebone connected to the thighbone—and all these bones are connected, somehow, to the story of how mankind grew—as we keep trying to piece it together

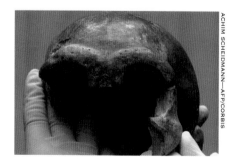

ACHIM SCHEIDMANN—AFP/CORBIS

Homo neanderthalensis

A skull unearthed in Germany's Neander Valley in 1856 belonged to a prehominid with a sloping forehead, a receding chin and thick ridges over his eye sockets. Above, a new Neandertals (now the favored spelling) skull was found in 2000. This early, small-brained species died out some 30,000 years ago, supplanted by *Homo sapiens,* most scientists believe.

JON READER—SPL/PHOTO RESEARCHERS

Australopithecus africanus

South African anthropologist Raymond Dart, left, uncovered a skull belonging to a species of ancient man that was startlingly different from those then known, Neandertals Man and Java Man. He named the short (just over 3 ft. tall) upright-walker *Australopithecus africanus,* the southern ape of Africa. In 1938 Robert Broom, a fellow South African, uncovered an adult specimen of a heavier, larger type of *Australopithecus,* which he initially christened *Paranthropus robustus,* above.

Homo habilis

In 1972 Bernard Negeneo, a Kenyan member of Richard Leakey's fossil-hunting team, spotted a few scraps of bone exposed by erosion in sandy sediments in a steep gully near Lake Turkana in East Africa. The dig that followed unearthed the skull and fragments at right, known as 1470, after their museum number. These are examples of the prehominid called *Homo habilis*, a toolmaking creature who lived some 2 million years ago. Many scientists believe "handy man" was a direct precursor of mankind.

Peking Man

During the 1920s, a series of digs in a cave at Choukoutien, China (above, a recent dig at the site), provided evidence of a new kind of prehominid, dubbed Peking Man, an example of the relatively recent species *Homo erectus*. The teeth, nearly complete skullcap and bone fragments found (inset, at right) have been dated between 750,000 to 500,000 years old.

Attention would-be screenwriters: the remains of Peking Man were turned over to U.S. Marines in China shortly after the Japanese attack on Pearl Harbor in 1941 and then mysteriously disappeared. They were never recovered.

Lucy

Lucy, perhaps the most celebrated find of the 20th century, was unearthed in 1974 by anthropologists Donald Johanson of the U.S., left, and Maurice Taieb of France—and yes, she was named after a Beatles song that was playing on the radio the day she was found. The best-preserved example of *Australophithecus afarensis* ever found, Lucy stood only 3½ ft. tall (males were 5 ft. tall) and had the brain capacity of a chimp. She is also getting on a bit—at some 3 million years old.

One Scientist's Evolution

Whether reimagining evolution, doing battle with creationists or writing his provocative essays, Stephen Jay Gould delights in upsetting our equilibrium

HARVARD PALEONTOLOGIST STEPHEN Jay Gould is a scientist blessed with a gift for explication: he excels at illuminating the ins and outs of his favorite subject, evolution, often by linking an observation of the smallest fossil to a grand, unexpected conclusion about the nature of life. He rose to prominence in his field in 1972, when he teamed up with Niles Eldredge of the American Museum of Natural History to propose the theory of punctuated equilibrium, which posited that evolution occurs by fits and starts, rather than by the gradual process proposed by the traditional view of evolution.

In recent years Gould has assumed a role as a spokesman against creationism in the schoolroom. When the Kansas state board of education mandated in 1999 that creationism be taught in the state's science classes, Gould told a University of Kansas audience, "[It's] like saying we're going to continue teaching the English language, but grammar is optional." Gould is well prepared for a long struggle: he's a New York Yankees fan who lives in Boston. Below are excerpts from a TIME interview with the paleontologist:

Q: *You have written that humankind is an afterthought, a cosmic accident. Why?*
A: Only in the sense that every species is. Since evolution has no inherent or predictable direction, if you could play life's tape again from any early point, you would get a completely different result that wouldn't include human beings. In that sense, every species' appearance is not random, because after it happens it is perfectly explainable, but it's unpredictable.

The reason I call human beings even more of an afterthought than others is that our lineage is so young and so small. The splitting point between human ancestors and those that gave rise to chimps and gorillas is 6 million to 8 million years ago, and the human species, *Homo sapiens*, is probably only about a quarter of a million years old.

Q: *So the view of evolution as a ladder with humankind on the top rung is incorrect?*
A: It is nothing more than a representation of our hopes. We have certain hopes and cultural traditions in the West, and we impose them upon the actual working of the world.

Q: *Does extinction mean failure?*
A: Extinction is the fate of all creatures ultimately. That's why it is so arrogant of us to think of dinosaurs as unsuccessful because they are dead. After all, they were around for 120 million years or so, and we have been around for only 250,000. And what's the chance that we're going to live for 500 times longer than we have already?

Q: *Has human progress led us to a point where technology might cause our own extinction?*
A: I think that is why our prospects for survival are really not great. People talk about human intelligence as the greatest adaptation in the history of the planet. It is an amazing and marvelous thing, but in evolutionary terms, it is as likely to do us in as to help us along.

Q: *Why is your work so popular?*
A: It's the subject more than anything else. I often say there are about half a dozen scientific subjects that are immensely intriguing to people because they deal with fundamental issues that disturb us and cause us to wonder. Evolution is one of those subjects. It attempts, insofar as science can, to answer the questions of what our life means, and why we are here, and where we came from, and who we are related to, and what has happened through time, and what has been the history of this planet. These are questions that all thinking people have to ponder.

THE ESSENTIAL GOULD

THE PANDA'S THUMB (1980)
Illustrates the ways in which natural imperfections and leftovers contribute to the evidence for evolution

THE MISMEASURE OF MAN (1981)
Why those who use statistical measurements often draw the wrong conclusions from the data

HEN'S TEETH AND HORSE'S TOES (1983)
The ingenuity by which fundamentally inappropriate body parts are pressed into new roles, like toes that become hooves

THE FLAMINGO'S SMILE (1985)
Posits that science and ethics are related yet ultimately independent

FULL HOUSE: THE SPREAD OF EXCELLENCE FROM PLATO TO DARWIN (1996)
Argues that human beings are neither the end nor the purpose of evolution

ROCKS OF AGES: SCIENCE AND RELIGION IN THE FULLNESS OF LIFE (1999)
Arguments against creationism, but not religion

MAMMOTHS: FIVE QUESTIONS

HOW BIG WERE THEY?
Jarkov, the mammoth discovered in Siberia, weighed six tons and stood 11 ft. tall. Some mammoths reached 10 tons and 14 ft. in height at the shoulder

WHEN DID THEY LIVE?
Mammoths roamed the earth in large numbers from about 2 million to 9,000 years ago. Jarkov, a 47-year-old male, is thought to have died 20,380 years ago. The last holdouts died on Wrangel island in the Bering Strait about 4,000 years ago, at about the time Stonehenge and the pyramids were built

WHERE DID THEY LIVE?
Mammoths first lived in Africa but eventually migrated to Europe and Siberia, then crossed the land bridge to Alaska and roamed North and South America

WHY DID THEY DIE OUT?
Theories fall into three categories: overkill, overchill or over-ill (i.e., early human predation, a quick change in climate or an as-yet-unknown virus)

ARE THEY EARLY ELEPHANTS?
No, but the two mammalian species probably shared a common ancestor

Wool Gathering

Genardi Jarkov just wasn't having any luck. In 1997 the nine-year-old boy was looking for reindeer on the Siberian plain where his nomadic family had herded the beasts for generations. He didn't find any reindeer that day, but, as it turned out, Genardi didn't go home empty handed either. After stumbling over what looked like a rock with hair growing out of it, he recognized an ivory tusk sticking up from the frozen ground. Knowing that full-sized tusks could earn his clan thousands of dollars, Genardi led his family back to the site. They promptly sawed off both tusks and set about trying to sell them.

After a circuitous journey, the tusks eventually came to the attention of French explorer Bernard Buigues, who realized that they belonged to a woolly mammoth, a massive, elephant-like species, most of whom died out roughly 10,000 years ago. Buigues hurried to the site and quickly determined that the best-preserved mammoth specimen ever found (even its hair was intact) was buried in the permafrost. Research showed the beast had frozen solid moments after it keeled over.

Hoping to keep Jarkov (as the beast was named) in pristine condition, Buigues and his team excavated a 23-ton cube of permafrost surrounding it and, in November 1999, flew it by helicopter to an ice cave more than 200 miles away, where they began painstakingly defrosting it with hair dryers (1½ years later, only 1% of Jarkov had been thawed out). While *Jurassic Park* fans await news that the animal has been successfully cloned (a very long shot, at best), scientists are enjoying a windfall of less dramatic but more important developments, like the discovery of ancient pollen grains and now extinct insects caught in the beast's hair—and blades of prehistoric grass wedged under its feet.

TUSKBUSTERS Top left: a cache of some 100 mammoth tusks found on Siberia's Taimyr Peninsula. Bottom left: Bernard Buigues dries out a swatch of hair in which ancient pollen (inset) was found. Bottom right: the huge cube of permafrost containing Jarkov was excavated; the tusks are a whimsical touch

1995: A NEW LASCAUX

When the French government announced in 1995 that a local official, Jean-Marie Chauvet, had discovered a stunning trove of Paleolithic art in a cave near Avignon, experts swiftly hailed the 30,000-year-old paintings as a cache rivaling—and perhaps surpassing—the two most famous sites of such art, Lascaux in France and Altamira in Spain.

The paintings and engravings, more than 300 of them, amount to a sort of Ice Age Noah's ark—images of bison, mammoths and woolly rhinos, of a panther, an owl, even a hyena. Done on the rock walls with plain earth pigments—red, black, ocher—they are of singular vitality and power.

How were such images painted? The main technique of Cro-Magnon art, according to pre-historian Michel Lorblanchet, director of France's National Center of Scientific Research, involved not brushes but a kind of oral spray painting: blowing pigment dissolved in saliva on the wall. Lorblanchet, testing his theory, has managed to re-create cave paintings with uncanny accuracy.

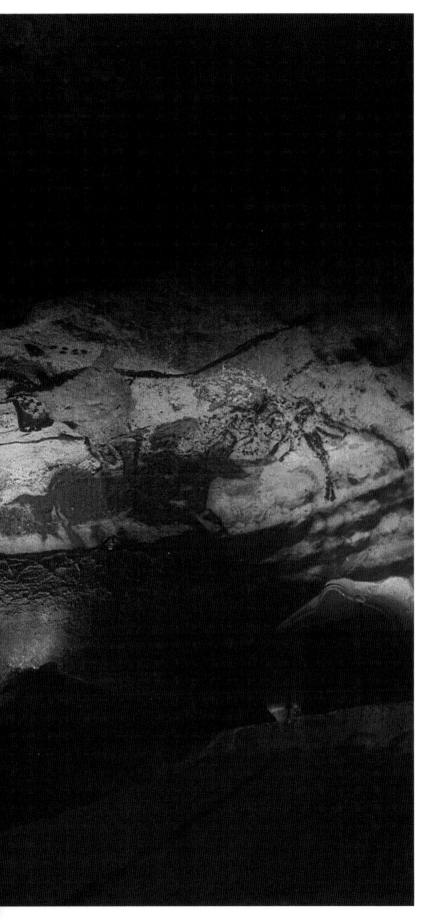

Vibrant Art from An Age of Stone

It's one of the most enduring stories of un-witting discovery: four young French boys out on a lark in September 1940, near Lascaux in the Dordogne region, decide to explore a hole in the ground one of them has recently found. They toss stones in the hole, determine it is deep and set out to explore it. Once inside, they are amazed to see gigantic, vivid paintings of animals covering the walls. The four boys bring their schoolteacher back to the cave; stunned, he contacts Abbé Henri Breuil, France's foremost prehistorian, who pronounces them authentic. Already, word of mouth is bringing tourists to the cave. There they are dazzled to find cavern after cavern covered with paintings—some 200 painted and drawn animals and symbols, along with nearly 1,500 engravings.

This splendid bestiary still has the power to amaze—although it was almost lost owing to careless maintenance. Over the years, as the cave was opened to tourists, the paintings gradually faded under artificial light, then were invaded by algae and bacteria. The cave was closed to visitors in 1963; today's tourists visit a replica.

When TIME first reported the discovery of the paintings in 1941, it cited French authorities in dating them at 30,000 B.C. More recently, radio-carbon dating has established their date as 15,000 B.C. They are the work of early men called the Magdalenians, after La Madeleine, the site of a rock-center shelter where signs of their culture were first found. The artists apparently stood on scaffolding to reach the ceiling; holes for wooden poles have been found in the walls. The painters mixed their pigments on the spot, and mixing tools have been found at other nearby sites. In all, there are some 200 caves in southwest Europe alone featuring Paleolithic art; the caverns at Lascaux remain the most impressive.

Photograph by Sisse Brimberg and Norbert Aujoulat—National Geographic Society

HALL OF THE BULLS The most spectacular cavern at Lascaux features these giant bulls; the largest one is more than 18 ft. long. The "canvas"—the cavern walls—consists of white calcite, perfect for holding pigment

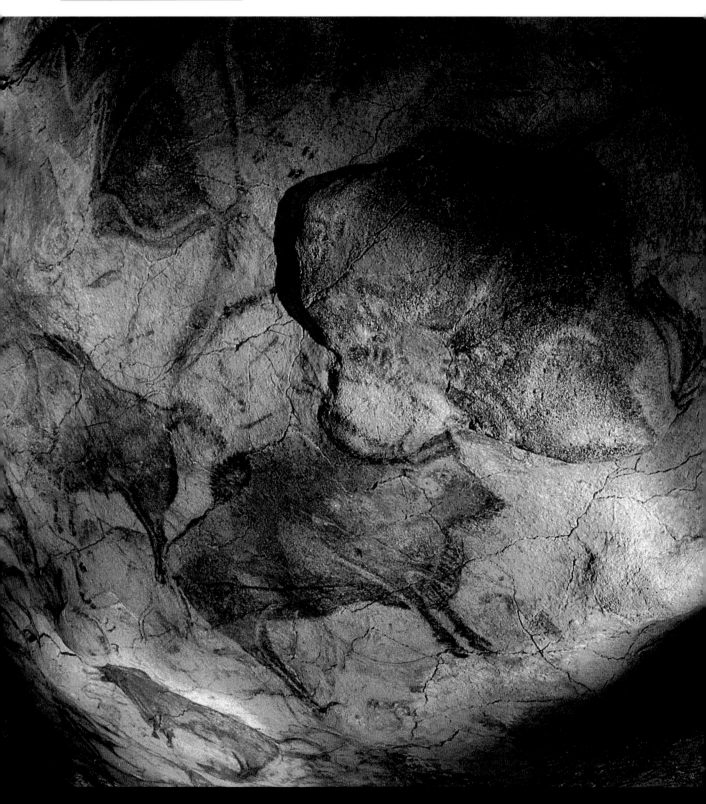

Prehistory's Picassos

These pictures, painted many thousands of years ago on the walls of caves, offer us our closest encounter with the minds of early men. Yet even scientists are divided on the function they served: Were they a hunting people's form of sympathetic magic? Or were they simply art for art's sake? However remote their purpose, their power is undeniable.

Lyrical Lascaux

In contrast to the earlier paintings found near Avignon, which feature dangerous animals like cave bears, panthers and woolly rhinos, those at Lascaux feature nonthreatening beasts, like the deer above. The animals were painted with a lyrical sweep and freshness that have survived the passage of 17,000 years.

The Cave at Altamira

Civilization first awoke to the splendor of prehistoric paintings late in the 19th century, when the magnificent works on the vault of a cave in Altamira, Spain, were discovered. Maria, the daughter of nobleman Marcelino de Sautuola, first noticed the paintings in 1879 on the ceilings of the cave; her father had previously found animal bones and flint tools on the site. Sautuola published an account of his findings in 1880, but the works were not recognized as genuinely prehistoric until the 20th century.

The main chamber that contains most of the paintings at Altamira measures about 60 ft. by 30 ft.; its roof is covered with paintings, chiefly of bison. There are also horses, wild boars and other figures, as well as handprints and hand outlines. The work has been dated to 12,000 B.C.

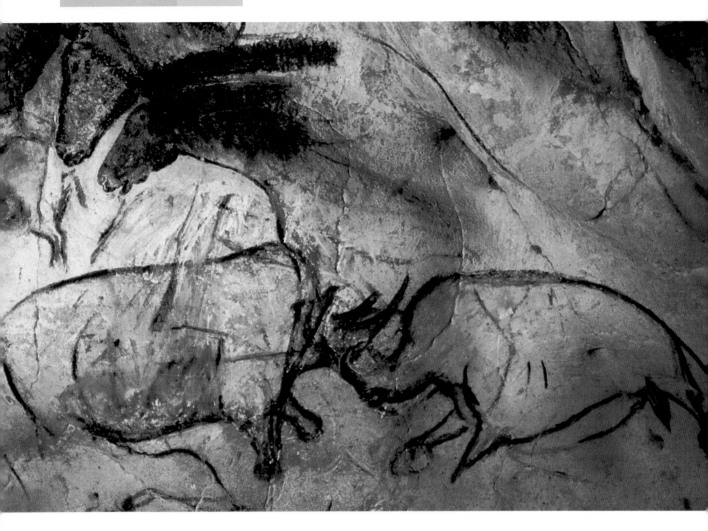

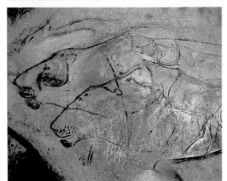
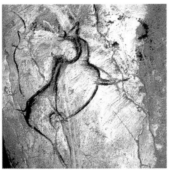
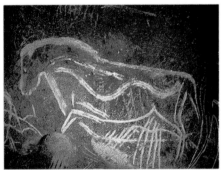
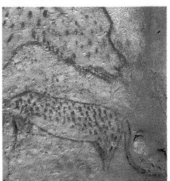

Woolly Rhino Battle

These seven pictures are all from the cave found near Avignon in 1995 by Jean-Marie Chauvet. "This is a virgin site — it's completely intact. It's great art," exulted Jean Clottes, a leading French authority on prehistoric art, soon after the cave was found. Another trove of ancient paintings, found in Cosquer, France, in the 1990s, can be reached only by scuba divers: its entrance lies under the surface of the Mediterranean.

Paleolithic Menagerie

Clockwise from top left, the artists of Chauvet painted horses; a stag; a feline and hyena; and three lions. Also found in the cave near Avignon was a bear skull placed with evident care atop a slab of rock; fragments of many more skulls were discovered on the cavern floor nearby.

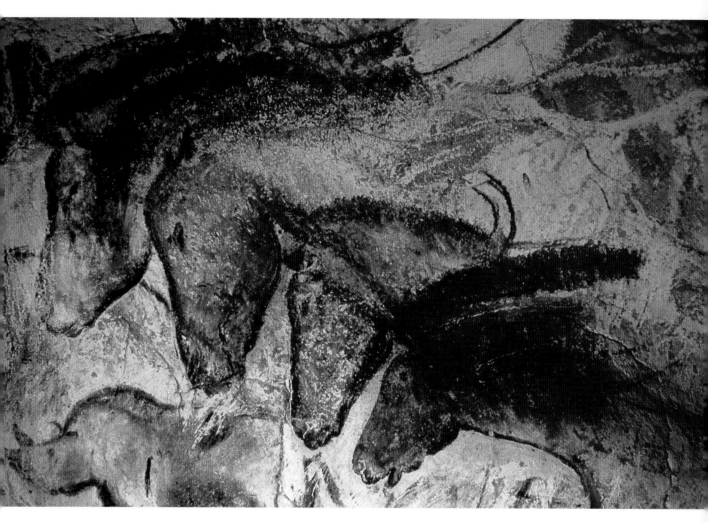

Running Across Time

The long-forgotten illustrator who created these images of horses had a fine eye for detail; despite popular myth, the Cro-Magnons were not simply inarticulate Alley Oops.

Maker's Mark?

Observing these images, TIME art critic Robert Hughes wondered, "Why the profuseness of Cro-Magnon art? Why did these people, of whom so little is known, need images so intensely? Why the preponderance of animals over human images? Archaeologists are not much closer to answering those questions than they were a half-century ago, when Lascaux was discovered." Whatever the answers, we can assume their creators were proud of their work—and one may have left his imprint, at right.

JEAN CLOTTES-FRENCH MINISTRY OF CULTURE AND COMMUNICATION, DRAC RHONE-ALPES, SRA

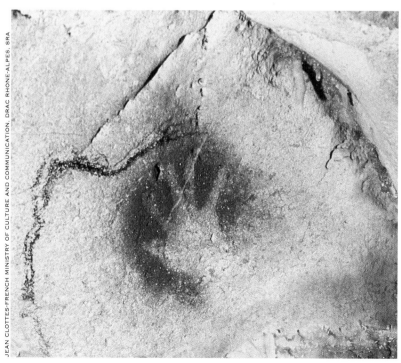

INSIDE A
STONE AGE
KIT BAG

TALISMAN?
Scientists are uncertain whether the polished stone circlet with a tassel was a necklace or fetish object

DAGGER
Fitted to an ash handle with straps, the flint blade was still razor sharp when found

CHOPPING TOOL
At first mistaken for bronze, the almost pure copper ax blade had hammered flangelike edges

ILLUSTRATION FOR TIME BY BURT SILVERMAN

The Iceman

Helmut Simon, a German tourist, first spotted the remarkably preserved remains of the Stone Age wanderer who came to be called the "Iceman" in the melting Similaun glacier high in the Alps in 1991. A comedy of errors followed: an Austrian policeman tried to pry the body from the ice with a jackhammer; curiosity seekers snitched fragments of his garments. Five days later, the find was finally brought to Konrad Spindler, head of the Innsbruck Institute for Prehistory, who said, "I thought this was perhaps what my colleague Howard Carter experienced when he opened the tomb of Tutankhamen and gazed into the face of the Pharaoh."

Who was the Iceman? Radio-carbon dating established his age at approximately 5,300 years, by far the most ancient human being ever found virtually intact. He stood 5 ft. 2 in. tall—short even in his day—and weighed around 110 lbs. Well prepared for the Alpine chill, he wore an unlined fur robe, whip-stitched together in a mosaic-like pattern, under a woven grass cape. His shoes were made of leather; his ax blade was nearly pure copper. He bore a fur quiver laden with a dozen incomplete arrows and two primed for shooting, with flint points and feathers. His bow was made of yew. He also was armed with a tiny, wooden-handled flint dagger; he carried a net of grass and a stone-and-linden tool probably used to sharpen points.

Yet for all his sophisticated gear, the Neolithic wanderer was behind the times: while his mountain people still hunted and gathered, far more advanced civilizations were flourishing elsewhere. In the Iceman's day, Alpine Europe lagged far behind Africa, the Middle East and South America in agriculture, commerce and transportation; in Sumeria men had just discovered the wheel.

Photograph by Paul Hanney—Gamma-Liaison

MOUNTAIN MAN Though the Alpine weather 5,000 years ago was milder than today's, experts believe the Iceman was in a state of exhaustion when he died, perhaps as a result of being caught by a cold front. Blue lines on his legs revealed that humans began tattooing their bodies earlier than previously believed

HIS 15 MINUTES CAME LATE This rendering is based on artifacts found near the frozen body. The Iceman's hair had been hand cut, proving that humans were barbering some 2,500 years before scientists had thought

Crete's Heroic Age

It seems incredible that one man could be responsible for opening our eyes to an entire culture, but until British archaeologist Arthur Evans successfully excavated the ruins of the palace of Knossos on the island of Crete, the great Minoan culture of the Mediterranean was more legend than fact. Indeed its most famed denizen was a creature of mythology: the half-man, half-bull Minotaur, said to have lived in a labyrinth under the palace of mythical King Minos. But as Evans proved, this realm was no myth. In a series of excavations in the early years of the 20th century, Evans found a trove of artifacts from the Minoan age, which reached its height from 1900 to 1450 B.C.: jewelry, carvings, pottery, altars shaped like bull's horns, and wall paintings showing Minoan life.

Evans also found what may have served as the model for the labyrinth itself, the many-corridored, 1,500-room palace of Knossos, near today's port city of Heraklion. Though modern-day archaeology has no place for the sort of flamboyant reconstruction Evans performed on the palace, with its rebuilt and painted pillars, the Heraklion tourist business remains in his debt. A 2000 book, *Minotaur: Sir Arthur Evans and the Archaeology of the Minoan Myth* by J. Alexander MacGillivray, skewered Evans as an unscientific, self-promoting amateur not unlike Heinrich Schliemann, whose 19th century excavation of the site of ancient Troy became the template for bad archaeology.

One riddle remains unanswered: the reason for the demise of Minoan culture, a fall so sudden it is believed to have involved a catastrophe. The eruption of the volcano on the nearby island of Santorini was long suspected as the cause, but that event is now dated at around 1625 B.C. Major Minoan sites were first destroyed about 1700 B.C., scientists now believe, with a second sudden cataclysm that doomed the civilization about 1450 B.C. The fall of Minoan Crete remains one of the great mysteries of history.

Photograph by Noboru Komine—Photo Researchers

MINOAN GLORY Murals from the palace of Knossos display classic Mediterranean motifs: from left, porpoises cavort, maidens bear amphoras, and acrobats leap over bulls. The "bull dancers" scene may relate to the myth of the Minotaur; bulls were sacred in Minoan life

IN EGYPT, TWO MORE FINDS

Above, King Tut is being anointed by his mother, the Queen, in a painted golden inlay from the back of his throne. Howard Carter's discovery of Tut's tomb remains unparalleled. But two sites discovered in the 1990s—seen on the following pages—have yielded fresh insights into ancient Egypt.

In 1995 Kent Weeks, an Egyptologist with the American University in Cairo, announced that in 1988 he had pried open a door blocked for thousands of years in the Valley of the Kings—not 200 ft. from Tut's tomb. There he found the most complex tomb ever found in Egypt, the resting place of the sons of Ramesses II, the ruler believed to have been Moses' nemesis in the *Book of Exodus*.

Since 1996, Zahi Hawass, a leading Egyptian archaeologist, has been exploring tombs in Bahariya, 230 miles southwest of Cairo, which are of two eras: the 26th dynasty (6th century B.C.) and the 1st and 2nd centuries A.D., when Egypt was a province of the Roman Empire.

A Boy King's Wondrous Things

Time was running out: the British archaeologist's aristocratic patron, Lord Carnarvon, had grown weary of bankrolling almost 20 years of fruitless digging. And the experts declared there was nothing left to be found here: after 200 years of excavations that had uncovered more than 60 tombs, Egypt's Valley of the Kings had given up all its secrets. And who was this Tutankhamen, anyway? An obscure boy king from the 18th dynasty about whom almost nothing was known—except that he didn't rate more than a footnote in the history of ancient Egypt.

Even so, Howard Carter—an illustrator by profession, an archaeologist by passion—couldn't let go of the idea that Tut's tomb was waiting to be found. So when his army of 200 diggers uncovered the top of a staircase on the morning of Nov. 4, 1922, he was sure that this was what he'd been waiting for. It was. When Carter first broke through the outer door, Carnarvon, summoned hurriedly from England, asked impatiently, "Can you see anything?" "Yes, wonderful things," Carter answered, transfixed.

What Carter saw was the first Egyptian royal tomb ever to be discovered intact. All the others found, before and since, had been looted by grave robbers thousands of years ago. Tutankhamen's tomb was spared precisely because of his obscurity; it was buried beneath that of a more prominent successor.

It took Carter and his team more than a decade to excavate and catalog the artifacts in Tut's tomb. When the job was done, there were more than 5,000 pieces of pharaonic treasure, including a 200-lb. solid-gold coffin and what has become perhaps the most photographed mask in the world: the face of a boy king, etched into beaten gold, that no one but Howard Carter thought mattered enough to search for.

Photograph: Archive/Liaison

GOLDEN SLUMBERS Carter working with Tut's gold coffin. His discovery sparked a rage of "Tut-mania" in the 1920s: faux-Egyptian fashions and styles were wildly popular

The Boy in the Golden Mask

Strange animals, statues and gold—everywhere the glint of gold," Howard Carter declared of Tutankhamen's tomb. The inlaid golden mask that covered the head of Tut's mummy bears enduring witness to the riches—both material and artistic—of ancient Egyptian culture. After 33 centuries, it is alive with vitality and intelligence

Journey to the Next World

The mural behind Tut's sarcophagus depicts his passage through death on his journey to becoming a god. The figure on the extreme left (wrapped in white) is Osiris, god of death. The next figure is Tut (now dead). The third figure is Tut in the form of his Ka (life force) commending his dead body to Osiris. The fourth figure is Nut, female goddess of the sky, greeting the living form of Tut (the fifth figure) as he ascends to his place alongside the gods. The sixth figure (obscured) is Tut transfigured as Osiris.

Ancient Gem

Tut's mask, left, is made of solid gold. It is 21⅛ in. high and weighs 22 lbs. The inlays are of turquoise; the stripes are lapis lazuli. The eyes are made of quartz and obsidian.

The Hunt

The statue at right portrays Tutankhamen standing on a papyrus skiff ready to harpoon an invisible prey—probably a crocodile. Once it is hit, the King will tie up his catch with the coiled rope held in his left hand.

A Family Affair

History knows him as Ramesses II, but to Egyptians he is Ramesses-al-Akbar—
Ramesses the Greatest. During his 67 years on the throne, from 1279 B.C. to 1212 B.C.,
he built more temples, obelisks and monuments, married more women (eight)
and claimed to have sired more children (162, by some accounts) than any other
Pharaoh. His empire stretched from present-day Libya to Iraq in the east, as far north
as Turkey and southward into the Sudan. The 1988 discovery and subsequent excavation
of his sons' mausoleum, above, offered new insights into the riches of his reign

Robbed!

Unlike King Tut's tomb, the mausoleum of Ramesses II's sons had been looted by grave robbers centuries ago: a 3,000-year-old papyrus fragment now housed in Italy recounts the trial of a thief who confessed under torture that he had broken into the tomb. Thousands of artifacts littered its floor, including beads, fragments of jars that were used to store the organs of the deceased, and mummified body parts. The tomb is immense: on both sides of this corridor are 10 doors, each leading to a 10-ft. by 10-ft. chamber; two additional corridors branch off, with 16 doors in each. On the far wall is a statue of Osiris whose face has been lost.

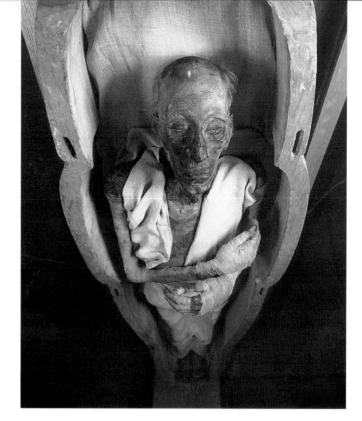

"Look on My Works, Ye Mighty, and Despair!"

Ramesses II's mummified body can be seen at Cairo's Egyptian Museum; it was interred in his tomb in the Valley of the Kings, which lies just 100 ft. away from the mausoleum that holds the bodies of his sons. Ramesses II was the model for Shelley's *Ozymandias.*

Hail to the Queen

Ramesses II's favorite wife, Queen Nefertari, sits in the center of the mural above, which was found in her tomb in the Valley of the Queens near Thebes in 1904. The deteriorating murals were closed to the public in 1940 but restored over six years beginning in the late 1980s.

Face of the Past

This pottery mask portrays a curly-haired Bedouin woman. Seven tombs from the 1st and 2nd centuries A.D. yielded copper bracelets, obsidian decorations and what may even be gaming pieces and dice.

A Donkey's Find

Egyptian archaeologist Zahi Hawass has been exploring an ancient necropolis in the city of Bahariya since 1996, when a donkey fell into a hole that led to an undiscovered tomb filled with gold-covered mummies. The tombs include that of Zed-Khonsu-efankh, the governor of the oasis in the 6th century B.C., as well as many from the 1st and 2nd centuries A.D.

At Death's Door

At this ornate tomb entrance from the Roman period, a guide leads the dead to the afterlife. Elsewhere, Hawass and his team found inscriptions representing the Book of the Dead. Hawass's *Valley of the Golden Mummies* (Abrams; 2000) is a richly illustrated guide to his excavations; the artifacts on these pages were unearthed after it was published.

Mass Grave

Families were often buried together; the site above contained 41 mummies. Hawass estimates that more than 10,000 mummies may be buried in the ancient cemetery. Thoroughly excavating the area, he estimates, may take at least 50 years—an archaeologist's dream.

A Charmed Life

These artifacts were found in the mausoleum of Zed-Khonsu-efankh, governor of Bahariya Oasis during the 26th dynasty. Though powerful, he was no King Tut; his sarcophagus was made of plain limestone. The gold and faïence amulets, depicting scarabs, snakes and gods, protected the governor on his journey into the underworld.

Goatherd's Find

Mohammed Adh Dhib Issa thought he knew where the goat was hiding. It was the winter of 1947, and the young Bedouin goatherd scrambled up to the mouth of a cave in the Qumran region of the Jordan River valley, overlooking the Dead Sea. He tossed a rock inside, hoping to trick the animal into giving itself away. But instead of a goat's bleating, Issa heard the sound of pottery smashing. When he investigated, the boy found seven leather scrolls wrapped in linen and stuffed inside terra cotta jars. Issa didn't know it, but the world had just changed.

The Dead Sea scrolls are the single most important archaeological find of the 20th century. Unearthed from 11 caves near Qumran and four additional sites in the Jordan River valley over the next 15 years, the scrolls contained, among other documents, a complete text of the Old Testament (with the exception of the *Book of Esther*, which was added to the Bible after the scrolls were written) that is more than 1,000 years older than the next oldest known manuscript, many additional religious texts and the mysterious Copper Scroll—purported to be a treasure map detailing the location of hundreds of pitchers full of gold, silver and jewels buried at more than 60 places in the desert near the Dead Sea (none of which has ever been found).

The scrolls themselves turned out to be a treasure for antiquities dealers (several of whom made fortunes on the discovery) and scholars (quite a few of whom built careers on translating and interpreting the texts). One person whose world didn't change very much, however, was Mohammed Adh Dhib Issa. He made less than $100 from his discovery. Today, just a single scroll—if you were able to buy one—would cost as much as $20 million. The goatherd died penniless.

Photograph by Graham Finlayson—Woodfin Camp & Assoc.

TIME CAPSULE This cave in Qumran, near the Dead Sea, sheltered the fragments of the ancient manuscripts for more than 20 centuries. At top right are portions of papyrus fragments; other manuscripts were written on leather

THE SCROLLS: FIVE QUESTIONS

WHEN WERE THEY FOUND?
The first batch of scrolls was found in 1947; further batches were found in the 1950s

WHERE WERE THEY FOUND?
The first discovery was at Qumran, on the northwest shore of the Dead Sea. Documents unearthed in four other sites, including the ancient fortress of Masada, are also designated Dead Sea scrolls

HOW OLD ARE THEY?
The documents date from the 3rd century B.C. to A.D. 68. They are written in Hebrew and Aramaic

WHO WROTE THEM?
Most scholars believe the scrolls were written by the Essenes, a monastic sect of Judaism; others believe they were not written by the Essenes, but were a library of works they protected from the Romans

WHY WERE THEY CONTROVERSIAL?
For decades, access to them was limited to a few designated scholars. Their monopoly was not broken until the '90s

An Eye in Space Finds a Lost City

The city of Ubar on the Arabian Peninsula was one of the great trading centers of antiquity. But it was also Islam's Sodom, a place that God destroyed because of its wickedness. For centuries, Ubar was forgotten. T.E. Lawrence dreamed of locating the lost city, which Lawrence of Arabia called "the Atlantis of the sands," but he did not live to carry out the search. The city was found by a pair of archaeological amateurs—with a strong assist from space-age gadgetry.

Their quest began in 1982, when Emmy-winning documentary filmmaker Nicholas Clapp happened upon an explorer's evidence of an ancient road that once led to Ubar. Clapp teamed up with lawyer George Hedges to raise money and organize an expedition. For guidance, they turned to Space Imaging Radar, carried on space shuttles to peer underneath the deserts of Egypt and locate ancient riverbeds.

After initial skepticism, Caltech's Jet Propulsion Laboratory agreed to have astronauts take SIR photos in 1984 during two passes over southern Arabia by the shuttle *Challenger.* Sure enough: the images showed ancient caravan trails beneath the sands. The routes converged at an area marked Omanum Emporium (the Omani Marketplace) on a map drawn by Ptolemy in the 2nd century A.D. (opposite page). The spot is in present-day Oman at the edge of the Empty Quarter, a trackless region infested with camel spiders, giant ticks and lethal carpet vipers. On the ground the team followed the route, but searched without success. Finally, they decided to examine Ash Shisar, a water hole with ruins of a primitive fort. Using ground-penetrating radar and sounding devices, the explorers discovered extensive ruins underneath. Later digging uncovered an octagonal castle with high walls and towers, just as described in the Koran, that might have reached a height of 30 ft. Ubar was found.

Photograph: NASA

SANDS OF TIME Roads and rivers that are barely visible to explorers on the ground appear in images captured from hundreds of miles up in space; in this digital composite, the faint white lines trace long-abandoned caravan routes, some running underneath sand dunes that have grown to be 600 ft. high

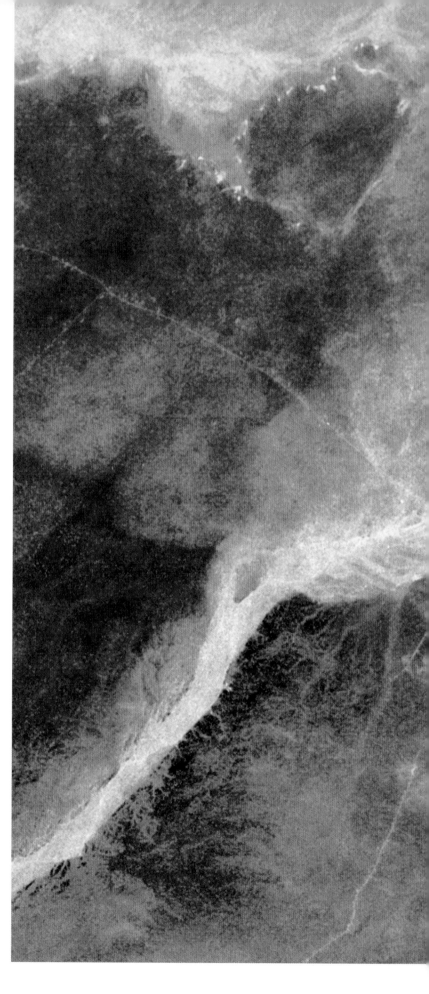

THE QUEEN OF SHEBA'S HOME

In all ancient Arabia, the most fabled land was the city of Ubar; above, it is "Omanum Emporium" on a map made by Claudius Ptolemy circa A.D. 150. As legend had it, one Shaddad ibn Ad created a jewel-encrusted oasis town in the southern deserts to stand as an "imitation of Paradise." Islam's holy Koran evoked the grandeur of "lofty pillars, the like of which were not produced in [all] the land" —before it succumbed to vice.

The area's ancient wealth was built upon its unique product, frankincense, the crystallized tree sap that was used as medicine, perfume and a preparation in cremation and embalming. Since this is the region whence the biblical Queen of Sheba made her trade mission to King Solomon, it is possible that frankincense from Ubar was burned regularly in the Jerusalem Temple. It is even conceivable that this oasis in the sands could have been the source of the frankincense that the Magi brought to the infant Jesus.

ALL THE KING'S MEN

The army of statues at Xian was commissioned to guard the royal tomb of Qin Shihuang, the emperor who defeated six warring states in 221 B.C. and became the first ruler to unify China, then set out to insulate his kingdom from barbarians by ordering the construction of its Great Wall. After making China one politically, the industrious leader unified it scientifically and culturally: he standardized the nation's weights and measures, its written language and its currency.

The great emperor's mausoleum—built by an estimated 700,000 workers over a period of 36 years—was put to the torch only four years after his death. Its ruins lie beneath a 250 ft. mound of earth near the four pits where the statues are being excavated. Other nearby remains awaiting excavation include the remnants of a palace and pits containing bronze chariots and the skeletons of horses and rare animals.

Buried Treasure

Part of the romance of archaeology is the sense that any one of us, at any time, might stumble upon a relic from a vanished past ... a hand-hewn stone arrowhead, a Civil War bullet. Indeed, some of the most significant discoveries in archaeology—from the cave paintings at Lascaux to the Dead Sea scrolls—have been made by amateurs. None surpass the magnificent army of warriors and horses unearthed in Xian, China, in 1974, by local farmers digging a well. So far the trove has yielded more than 7,500 life-size figures made of terra cotta. Most impressive of all, the figures were modeled from life, not cast in a mold; each soldier, each horse, is unique, detailed down to strands of hair and patterns on the soles of shoes.

The figures are arrayed in a series of four separate pits; tourists can observe the ongoing process of excavation and restoration from a series of exhibit halls. The statues are found broken into pieces, which are coded, cleaned and placed together, an elaborate jigsaw puzzle. Though their uniform clay color lends them classical simplicity, it is the work of time, not man. The original statues were brightly painted, using mineral-based pigments, including charcoal to color the hair. Some statues retain flakes of paint, clues that allow scientists to attempt restoring them to their original appearance.

When Chinese scientists first arrived at the site to oversee the excavation, they assumed their work would occupy them for several weeks at most. More than a quarter-century later, the work goes on: the number of statues still hidden underground far outpaces the number restored to date.

Photographs by O. Louis Mazzatenta—
National Geographic Society

GUARD OF HONOR The statues were originally sheltered from the elements by a roof made of pine logs, which scientists believe collapsed as a result of the fires set by rebels after the emperor's death

Maya Sunset

Who were the Maya, the people who built majestic cities, pyramids and statues all over Central America, from Yucatán to modern Honduras—and later abandoned them a few centuries before the Spanish conquistadors arrived? The question has piqued scientists since the ruins were first encountered and described in the 19th century.

Archaeologists have long known that the Maya, who flourished between about A.D. 250 and 900, perfected the most complex writing system in the hemisphere and mastered mathematics and astrological calendars of astonishing accuracy. Among the myths about them to be debunked in recent decades is that they were a peaceful race. Experts now generally agree that warfare played a key role in Maya culture. The rulers employed torture and human sacrifice on many religious and sporting occasions, even during building dedications.

Uncontrolled warfare was probably one of the main causes for the eventual downfall of the Maya. In the centuries after A.D. 250 —the start of what is called the Classic period of Maya civilization—the skirmishes that were common among competing city-states escalated into full-fledged, vicious wars that turned the proud cities into ghost towns. Overpopulation may have played a key role: too much exploitation of the rain-forest ecosystem, on which the Maya depended for food, as well as water shortages, seem to have contributed to the collapse.

Photograph by Robert Frerck — Woodfin Camp

SUNDIAL Each spring and fall equinox at El Castillo at Chichén Itzá on Mexico's Yucatán Peninsula, the setting sun's serrated shadow slowly moves up the staircase, whose base begins with two massive serpent heads. The pyramid was dedicated to Kukulcan, a potent god of creation and transformation. Its four staircases have 91 steps each, which, added to the top platform, equal 365, the number of days in the solar year.

TALKING HEADS

Information about the Maya has come not just from physical objects but also from the elaborate hieroglyphics they left behind, like the "wall of skulls," above, in Chichén Itzá. On a 1970 visit to Mexico, former University of South Alabama art teacher Linda Schele was mesmerized by the ruins at Palenque. Three years later, she was adept enough to join two other experts and decode the history of Palenque over two centuries in a mere 2½ hours—and to get it right.

How was this possible? Because decoding glyphs depends heavily on intuition. Said Schele: "These moments of clarity are extraordinary. The greatest thrills of my career came in those moments when the inscription becomes clear and we suddenly understand the humans who created this legacy for the first time."

Like most official records, glyphs may contain a healthy dose of propaganda (imagine trying to understand the Gulf War by reading Saddam Hussein's speeches). Even so, glyphs are a portal into the past.

GIANNI DAGLI—ORTI-CORBIS

Ebb Tide of a Culture

Tulum, on Mexico's coast, presents the contrasts of Maya culture: lavish buildings on a beautiful site, yet devoted to human sacrifice

Maya-mania is easy to explain, according to Arthur Demarest, an archaeologist who led a team of researchers in unearthing the ruins of Dos Pilas, a onetime major Maya metropolis in Guatemala: "You've got lost cities in the jungle, secret inscriptions that only a few people can read, tombs with treasures in them, and then the mystery of why it all collapsed."

Among the first modern Westerners to be captivated by the Maya were the American John Lloyd Stephens and English artist Frederick Catherwood, who in 1839 started to bushwhack their way into the Central American rain forest to gaze at the monumental ruins of Copán, Palenque, Uxmal and other Maya sites. The account Stephens published of his trek in 1841, *Incidents of Travel in Central America, Chiapas and Yucatán,* was an enormous popular success and sparked others to follow him and Catherwood into the jungle and into musty Spanish colo-

nial archives. Over the next half-century, researchers uncovered, among other things, the *Popol Vuh* (the sacred book of the Quiche Maya tribe) and the *Relacion de las Cosas de Yucatán,* an account of Maya culture during and immediately after the 16th century Spanish conquest written by the Roman Catholic Bishop Diego de Landa. By the 1890s, Alfred Maudslay, an English explorer, was compiling the first comprehensive catalog of Maya buildings, monuments and inscriptions, and the first excavations were under way.

Arthur Demarest divides the history of the region into two periods: before 761 and after. Before that year, he says, wars were well-orchestrated battles to seize dynastic power and procure royal captives for very public and ornate executions. But after 761, he notes, "wars led to wholesale destruction of property and people, reflecting a breakdown of [the] social order."

PINNACLE Maya ruins crown a cliff in Tulum, on the Yucatán coast. The religious site was built in the waning days of Maya civilization, around A.D. 1280, although an upright column dating from A.D. 564 was found here. Above is Frederick Catherwood's illustration of Tulum as he and John Lloyd Stephens found it during their pioneering exploration of Central America's ruins

ROBERT FRERCK—WOODFIN CAMP

Through Architecture to the Stars

El Caracol (the Snail), above, so called for its spiral interior staircase, is the observatory at Chichén Itzá. It was aligned to determine the location of the planet Venus on specific dates, as well as some 20 other such alignments. Dating from the 10th century, it is one of the last buildings constructed at the site.

Blood Offerings

A chacmool figure reclining in front of a Chichén Itzá temple bears on its chest a vessel for offerings to the Maya gods. These gifts may have included hearts ripped from the chests of sacrificial victims. Chacmools, which also reflect the influence of central Mexico's Toltec culture, are thought to represent dead warriors in homage to the gods.

ROBERT FRERCK—WOODFIN CAMP

Even Rome Wasn't Built in a Day

Behold the Pyramid of the Sorcerer at Uxmal, 80 miles east of Chichén Itzá, built between A.D. 600 and 900. There are 121 steps, each just 6 in. wide. Uxmal means "thrice built," though there is evidence of five separate building periods in this unusual (for the Maya) oval-shaped structure. Maya legend holds that the pyramid was built in a single night by a dwarf magician. Anthropologists believe the steep stairways played a part in the most important rituals of the Maya, those of human sacrifice: victims were decapitated and their heads rolled down the stairs.

Soaring City

Perched 7,710 ft. high amid the sun-splashed Andes, the ancient Inca city of Machu Picchu is certainly one of the world's most breath-taking sights. To further feed the fancy, as one U.S. website proprietor proclaims, tongue in cheek: It is Far Away and has a Strange-Sounding Name. The city was not detected by the Spaniards and was "discovered" (a loaded cultural term, in this case) by science on a 1911 expedition led by Yale University professor Hiram Bingham. Like the conquistadors centuries before him, Bingham was seeking a lost city—but not this one. He was on the track of Vilcabamba, the "lost city of the Incas," from which the Incas last, futile rebellion against the Spanish conquerors was launched. Bingham believed he had found it here; modern scholars do not agree.

The city was apparently constructed between the mid-15th to the early or mid-16th century. Its forbidding location—paired with the discovery of a sizable number of female skeletons at the site—has led many scientists to believe the city primarily served a religious purpose, perhaps as a sanctuary for the Virgins of the Sun, an élite Inca social group.

Peru's upkeep of the site has recently come under fire, particularly since authorities announced plans to open a cable-car system that would double tourist access to it—and since a crane collapsed during the filming of a beer commercial in 2000, damaging the priceless ruins. UNESCO declared Machu Picchu a World Heritage Site in 1983, and has said it is considering taking over control of the site to ensure its protection.

Photograph by Brian Vikander—Corbis

H. BINGHAM, YALE U.—NATIONAL GEOGRAPHIC SOCIETY

2001: A NEW FIND IN PERU

After Yale history professor Hiram Bingham, above, was led to the ruins of Machu Picchu by Melchor Arteaga, in 1911, the American returned to conduct excavations in 1912 and 1915.

The exploration goes on: In April 2001, ancient sites in Peru again made the news when scientists investigating Caral, a long-ignored Peruvian site, declared that it is the oldest city in the Americas, with a highly structured society that flourished at the time when the pyramids were being built in Egypt.

Caral is one of about a dozen large sites in the Supe Valley, just inland from the Pacific coast in central Peru. It was first uncovered in 1905 but has not been fully explored. Radiocarbon dating has shown that the city flourished for five centuries, starting about 2600 B.C., with irrigation systems and public architecture— signs of a society with strong, centralized leadership.

The research team included scientists from the Field Museum in Chicago, Northern Illinois University and Peru's Universidad Nacional Mayor de San Marcos.

HIGH-RISE The mountain's sloping sides were terraced for agriculture; an aqueduct system funneled precious water to the fields. The Incas worshipped the sun: The site includes several temples and a ceremonial sundial

Sacrificial Lambs

By modern standards, the hostile summit of Mount Llullaillaco, in Argentina's Andes, is no place for kids. But the ancient Inca saw things differently, and so it was that one day some 500 years ago, three children ascended the frigid, treacherous upper slopes of the 22,000-ft. peak. They would not return. Once at the summit, the children—two girls and a boy, between eight and 15 years old— were ritually sacrificed and entombed beneath 5 ft. of rocky rubble.

And there the story might have ended but for the work of Johan Reinhard, an archaeologist funded by the National Geographic Society, who scales the Andes in search of sacrificial remains. He had already located 15 bodies, including the famed ice maiden he found in 1995. But these three, whose discovery he announced in April 1999, are by far the most impressive. They were frozen solid within hours of their burial. Two of the bodies are almost perfectly preserved; the third was evidently damaged by lightning. The children's internal organs are not only intact but also still contain blood.

A wealth of artifacts was buried along with the bodies: 36 gold and silver statues, small woven bags, a ceramic vessel, leather sandals, a small llama figure and seashell necklaces. The head of one of the girls sports a plume of feathers and a golden mask. Bundles of food wrapped in alpaca skin indicated that the children came from the Incan social élite. But the real riches were within: the preserved bodies gave scientists an unprecedented look at Incan physiology.

Photograph by Johan Reinhard—*National Geographic*

FOOD FOR THOUGHT Scientists examined the children's stomachs to find out what they ate for their last meal, their organs for clues about their diet and their DNA to establish their relationship to modern-day ethnic groups

THE INCA
ICE MAIDEN

Four years before he
found the mummies
pictured at left,
Johan Reinhard and
his Peruvian assis-
tant Miguel Zárate
discovered the well-
preserved remains of
an adolescent girl on
Mount Ampato in
Peru, above. TIME
listed the discovery
as one of the 10
most important
scientific events of
1995. The kneeling
posture in which the
body was found
indicates that the
girl may have been
praying at the time
of her death, a clue
that she was perhaps
a human sacrifice.
The remains were
examined at Johns
Hopkins University
Medical Center
Hospital in 1996.
DNA testing showed
that the young
woman shared the
genetic profile of
modern-day
Panamanians, but
also that of people
from Taiwan and
Korea, supporting
the theory that the
first Americans may
have migrated from
Asia. When the
"Inca ice maiden"
was displayed at
National Geographic
headquarters in
Washington in
1996, Senator Joe
Lieberman quipped:
"in high school,
she dated Bob
Dole."

RAIDERS—
AND TRADERS

You guessed it—the ruthless buccaneers at right are not Vikings; they are modern-day sailors enjoying a voyage on a reconstructed Viking ship, *Islendingur*, as it leaves Reykjavik harbor in June, 2000. The ship followed in the path of Leif Eriksson's voyage to Greenland and America 1,000 years before. Thanks to the remains of a number of Viking boats disinterred from burial mounds in Norway, scientists believe that the wooden craft built by these Scandinavian seafarers were the best in Europe at their time. Sleek and streamlined, powered by both sails and oars, quick and highly maneuverable, the boats operated equally well in shallow water and on the open seas.

Often portrayed as ravagers, pagans and heathens, the Vikings were indeed raiders, but they were also traders whose economic network stretched from today's Iraq all the way to the Canadian Arctic. They were democrats who founded the world's oldest surviving parliament while Britain was still mired in feudalism. Master metalworkers, they fashioned exquisite jewelry from silver, gold and bronze.

In their sturdy craft, the Norse searched far and wide for goods they could not get at home: silk, glass, sword-quality steel, raw silver and silver coins that they could melt down and rework. In return they offered furs, grindstones, Baltic amber, walrus ivory, walrus hides and iron. They worshiped a pantheon of deities, three of whom—Odin, Thor and Freya—we recall every Wednesday, Thursday and Friday.

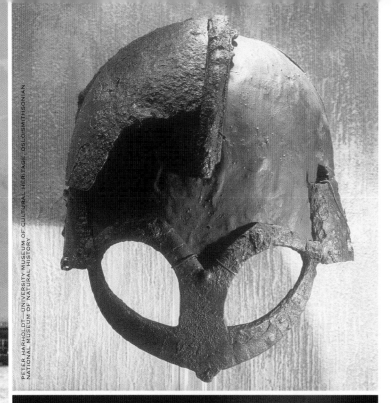

American Vikings

Not so many decades ago, the notion that the first European to voyage to North America was someone other than Christopher Columbus seemed … well, un-American. But thanks to the work of Norwegian explorer Helge Ingstad and his wife, archaeologist Anne Stine Ingstad, we now have physical evidence of the Scandinavian settlement of North America. In 1960 the couple went to Newfoundland to explore a place identified on an Icelandic map dating from the 1670s as "Promontorium Windlandiae," near the small fishing village of L'Anse aux Meadows, in the Canadian province's northern reaches. They were certain that it marked the location of an ancient Norse settlement.

Finding the place turned out to be absurdly easy. When the Ingstads asked the locals if there were any odd ruins in the area, they were taken to a place known as "the Indian camp." They immediately recognized the grass-covered ridges as Viking-era ruins, similar to those previously found in Iceland and Greenland. During the next seven years, the Ingstads and an international team of archaeologists exposed the foundations of eight separate buildings. Sitting on a narrow terrace between two bogs, the buildings had sod walls and peaked sod roofs laid over a (now decayed) wooden frame; they were evidently meant to be used year-round. The team also unearthed a Celtic-style bronze pin with a ring-shaped head similar to ones the Norse used to fasten their cloaks, a soapstone spindle whorl, a bit of bone needle, a small whetstone for sharpening scissors and needles, lumps of worked iron and iron boat nails—all not native to North America.

Further excavations in the 1970s made it plain that this was most likely the place where legendary Viking Leif Eriksson set up camp around A.D. 1000—just as stated in ancient Scandinavian sagas. Helge Instad, the man whose discovery rewrote the world's history books, died at age 101 in March 2001.

LODGES At far left below is the exterior of two restored structures at L'Anse aux Meadows; the interior of one is at near left. The settlement may have been used for only 10 years or so. At top left is the only known complete Viking helmet, found in Europe

Time Travel

Einstein's theory of relativity holds that the faster we travel, the farther (and faster) we move into the future. Thus an astronaut in a spacecraft moving at one-half the speed of light for one year would return to an earth on which thousands of years had elapsed. But as for the Hollywood fun stuff—like *Back to the Future* or this still from the 1960 film of H.G. Wells' *The Time Machine*—don't hold your breath.

Time Out!

We can't travel through time—yet. But that hasn't stopped a bevy of quacks, bunkum peddlers (and Hollywood studios) from concocting a collection case filled with frauds, hoaxes and what-ifs

The Bird-Dinosaur Link

In the fall of 1999 it was announced that *Archaeoraptor liaoningensis,* a meat-eating bird with a dinosaur-like tail, had been found in China. The remains of this "missing link" between bird and dinosaur were displayed at prestigious museums and featured in a *National Geographic* story. But the beast was a hoax staged by shady fossil traffickers in China. To its credit, *National Geo* investigated and published a quick retraction on "the Piltdown chicken."

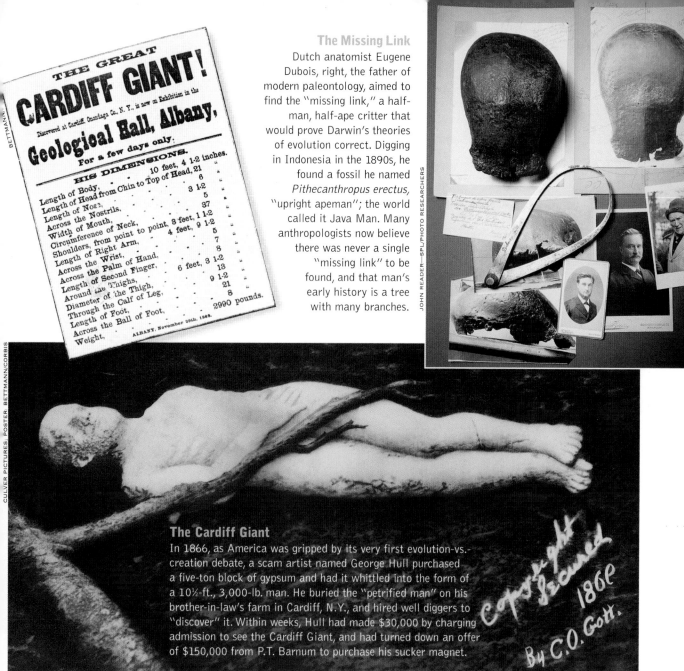

THE GREAT CARDIFF GIANT!

Discovered at Cardiff, Onondaga Co., N. Y., is now on Exhibition in the

Geological Hall, Albany,

For a few days only.

HIS DIMENSIONS.

Length of Body,	10 feet, 4 1-2 inches.
Length of Head from Chin to Top of Head,	21
Length of Nose,	6
Across the Nostrils,	3 1-2
Width of Mouth,	37
Circumference of Neck,	
Shoulders, from point to point,	3 feet, 1 1-2
Length of Right Arm,	4 feet, 9 1-2
Across the Wrist,	5
Across the Palm of Hand,	7
Length of Second Finger,	8
Around the Thighs,	6 feet, 3 1-2
Diameter of the Thigh,	13
Through the Calf of Leg,	9 1-2
Length of Foot,	21
Across the Ball of Foot,	8
Weight,	2990 pounds.

ALBANY, November 29th, 1868.

The Missing Link

Dutch anatomist Eugene Dubois, right, the father of modern paleontology, aimed to find the "missing link," a half-man, half-ape critter that would prove Darwin's theories of evolution correct. Digging in Indonesia in the 1890s, he found a fossil he named *Pithecanthropus erectus,* "upright apeman"; the world called it Java Man. Many anthropologists now believe there was never a single "missing link" to be found, and that man's early history is a tree with many branches.

The Cardiff Giant

In 1866, as America was gripped by its very first evolution-vs.-creation debate, a scam artist named George Hull purchased a five-ton block of gypsum and had it whittled into the form of a 10½-ft., 3,000-lb. man. He buried the "petrified man" on his brother-in-law's farm in Cardiff, N.Y., and hired well diggers to "discover" it. Within weeks, Hull had made $30,000 by charging admission to see the Cardiff Giant, and had turned down an offer of $150,000 from P.T. Barnum to purchase his sucker magnet.

Copyright secured 1869 By C.O. Gott.

Piltdown Man

When amateur geologist Charles Dawson, right center, uncovered ancient specimens with a skull that looked human but a jaw resembling a primate's at England's Piltdown Common in 1911, scientists went, well, ape. Doubts quickly surfaced, but it was only in 1953 that modern tests proved both specimens were much younger than first estimated—and that the jaw belonged to an orangutan. Dawson was apparently sincere; the author of the memorable hoax remains a mystery.

THE ILLUSTRATED LONDON NEWS.

SATURDAY, DECEMBER 28, 1912.

SIGNATURE IN THE SKY Comet Hale-Bopp was discovered as it traveled outside the orbit of Jupiter in 1995 by amateur astronomers Alan Hale and Thomas Bopp, working independently. Above, it soars over Mount Whitney in 1997

Photograph by Aaron Horowitz—Corbis

Discoveries in

Space

❝I don't feel rejected by the sky. I'm a part of it—tiny, to be sure, but everything is tiny compared to that overwhelming immensity. And when I concentrate on the stars, the planets and their motions, I have an irresistible sense of machinery, clockwork, elegant precision working on a scale that, however lofty our aspirations, dwarfs and humbles us. ❞ —Carl Sagan

The Rocket Scientist

Lampooned as a lunatic, Robert Goddard launched us into the Space Age, but he kept his work on missiles to himself—and a few interested German scientists

H E WAS A ROCKET MAN BEFORE HE WAS a man at all. Robert Goddard was born in 1882; from childhood, he had an instinctive feel for all things pyrotechnic and was intrigued by the infernal powders that fuel firecrackers and sticks of TNT. Figure out how to manage that chemical violence, he believed, and you could do some ripping-good flying. As a student and professor at Worcester Polytechnic Institute and later at Clark University, Goddard tried to find out just how. Fooling around with the arithmetic of propulsion, he calculated the energy-to-weight ratio of various fuels. Fooling around with chemistry, he learned that if he hoped to launch a missile very far, he would need something stronger than the black powder used in fireworks, something with propulsive oomph—a liquid like kerosene or liquid hydrogen, mixed with liquid oxygen to allow combustion to take place in the airless environment of space.

For nearly 20 years, Goddard's theories were just theories, and the rockets he built never flew out of Worcester—or anywhere at all. On Jan. 13, 1920, the New York Times ran an editorial lampooning a paper Goddard had published predicting that someday a rocket might reach the moon. Goddard never forgot the slight.

He got mad—then got even. On March 16, 1926, Goddard finished building a spindly, 10-ft. rocket he dubbed Nell, loaded it into an open car and trundled it out to his aunt Effie's nearby farm. He set up the missile in the field, then summoned an assistant, who lit its fuse with a blowtorch attached to a long stick. For an instant the rocket did nothing at all, then suddenly it leaped from the ground and screamed into the sky at 60 m.p.h. Climbing to an altitude of 41 ft., it arced over, plummeted earthward and

slammed into a frozen cabbage patch 184 ft. away. The entire flight lasted just 2½ sec.—but that was longer than any liquid-fueled rocket had flown before.

As Goddard's Nells grew steadily bigger, the town of Worcester caught on. In 1929, an 11-ft. missile caused such a stir the police were called. The next day the local paper ran a horse-laughing headline: MOON ROCKET MISSES TARGET BY 238,799½ MILES. For Goddard the East Coast was becoming cramped. In 1930 he and his wife Esther headed west to Roswell, N.M., where the land was vast, the launch weather good, and the locals minded their own business.

In the open, roasted stretches of the Western scrub, the fiercely private Goddard thrived. Over the next nine years, his Nells grew from 12 ft. to 18 ft., and their altitude climbed from 2,000 ft. to 9,000 ft. He built a rocket that exceeded the speed of sound and another with stabilized steering, and he filed dozens of patents for everything from gyroscopic guidance systems to multistage rockets.

Goddard often corresponded about rocketry with German scientists; when they fell silent in 1939, he showed U.S. Army officials films of his rockets and noted that they could be used as weapons. The Army brass thanked him and sent him on his way. Five years later, the first of Germany's murderous V-2 rockets blasted off for London. By 1945, more than 1,100 of them had rained down on the ruined city. When Goddard finally got a look at a V-2's innards after the war, he immediately saw the missile was based on his work. But before the year 1945 was out, he was dead of throat cancer. When men stepped on the moon in 1969, the New York Times wrote an editorial of apology to America's father of rocket science.

Photograph 1926—Smithsonian Institution

1882
Born Oct. 5 in Worcester, Mass.

1908
Begins studying physics at Clark University in Worcester

1915
Proves that rocket engines can produce thrust in a vacuum

1920
Derided in a New York Times editorial, he resolves to keep his research private

1926
Launches the first liquid-fueled rocket (Nell); it flies to an altitude of 41 ft.

1930
Begins working in Roswell, N.M., where he develops supersonic and multistage rockets and invents fin-guided steering

1945
Dies at age 62, holding more than 100 patents

1953
U.S. launches the first Redstone rocket, an evolution of Goddard's work

1969
U.S. Saturn rocket sends man on moon landing; New York Times apologizes to Goddard

Snowballs of Dirt

Everyone's favorite comet guru, Fred Whipple, has had two chances to see everyone's favorite comet, Halley's. But he blew the first visit, in 1910: "I was three years old," he explains. Like young Fred, astronomers at the time understood very little about comets. What, for example, were these solar orbiters made of? And why did their tails sometimes shoot out in front of them instead of trailing behind? Why were their orbits unpredictable? And why did their periods (the time between visits) always vary slightly? In 1910, the accepted answers were, respectively, "Flying sandbanks," "We don't know," "Just because," and "Stop asking annoying questions."

Four decades later, as Halley's comet was starting to circle back toward the sun, Whipple had come a long way from his Iowa boyhood and the world had come a long way in understanding comets. The second journey would not have been possible without the first. It was in 1950 that Whipple, now an astronomy professor at Harvard, first proposed his "dirty snowball" theory: comets consist almost entirely of frozen water vapor and gas (the "snowball" part) mixed with cosmic dust (the "dirty" part) that it picks up while traveling through the solar system.

This revolutionary idea illuminated comet behavior. The tail (being vapor and gas melted by the sun's heat and deflected by solar "wind") sometimes shoots out in front of the comet because it always points away from the sun—so as the comet recedes from the sun, it seems to be flying backward. The time between sightings of the same comet differs because these gaseous tails speed the comet up when behind it and slow it down when in front. And their orbits are eccentric because comets tumble: their jetlike "exhaust" fires in multiple directions, slightly varying its path.

And what about Whipple's second chance to see Halley's? In 1985 and '86, he had a front-row seat (as a NASA consultant) for its return—which proved less visually spectacular, if more scientifically rewarding, than other visits. Here's hoping the grand old man of comet studies, 94 in 2001, will stick around for a third gander at Halley's—in 2060.

Photograph by Jonathan Blair—Corbis

STAR CLUSTER Amateur astronomers set up their gear outside Tucson, Ariz., for a night of observing Halley's comet, center right, in 1986

DETOURS OF A COMET MAN

Fred Whipple may be the maven of comet studies, but he is also a scientist for all seasons. During World War II, he took time out from astronomy to invent antiradar chaff (the tiny strips of aluminum dropped by planes to confuse enemy tracking systems) for the Air Force. In the late 1940s he designed a "meteor bumper" for spacecraft that wouldn't be built until 20 years hence. Now called a "Whipple Shield," the bumper is standard equipment on all probes that leave Earth orbit.

In the '60s Whipple co-invented the Multiple Mirror Telescope, which used military surplus parts to focus light from six different reflective surfaces onto a single lens, thus creating an image that could otherwise be captured only by a much larger and more expensive telescope. He also directed the development of the first space-based telescope: the forerunner to the Hubble Space Telescope, the Orbiting Astronomical Observatory 1 was launched in 1968 but suffered a mechanical failure. Oh, yes: along the way, Whipple also found time to discover six previously unknown comets.

CLYDE
TOMBAUGH

Whether Pluto is
a planet or not, its
discovery is one of the
great stories of the
proof of a scientific
theory. For that's
what Pluto was in the
early years of the 20th
century—a theoreti-
cal body, somewhere
beyond the orbit of
Neptune. It was
astronomer Percival
Lowell, the man who
popularized the notion
of canals on Mars,
who posited Pluto's
existence, basing it on
his rigorous calcula-
tions of the orbit
of Uranus, which
showed gravitational
anomalies best
explained by the
presence of an as-yet-
undetected object.

Astronomer Clyde
Tombaugh set out to
prove Lowell correct.
In 1930, combing
through a series of
photographs he had
taken at the Lowell
Observatory in
Arizona, Tombaugh,
24, detected the new
planet by its motion,
which was much
slower than that of
numerous asteroids
also recorded on the
same pictures.

In 1978 two
Americans, James W.
Christy and Robert
S. Harrington,
discovered Pluto's
satellite Charon, an
object about half
Pluto's size. Some
12,200 miles separate
the two.

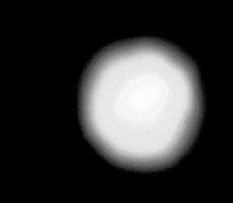

Is Pluto a Planet?

There's no doubt that Clyde Tombaugh discovered Pluto. But 71 years after his find, the question of what exactly Tombaugh discovered remains very much alive. When the new Hayden Planetarium at the $210 million Rose Center for Earth and Space at the American Museum of Natural History in New York City opened its doors in 2000, visitors were dazzled by its sphere-in-a-glass-cube architecture—so dazzled, it took a few weeks for an unsettling fact to sink in. The planetarium's soaring entry space included scale models of the planets of the solar system—all eight of them. Pluto had been excluded, banished, stripped of its planetary status.

Let the argument begin! "We're not that confrontational about it," said Neil de Grasse Tyson, director of the planetarium. "You actually have to pay attention to make note of this." Attention was paid. "They went too far in demoting Pluto, way beyond what the mainstream astronomers think," Richard P. Binzel, a professor of planetary science at the Massachusetts Institute of Technology, told the New York *Times*.

Pluto is certainly unique among the planets. Its composition is similar to that of a comet. Its orbit is elliptical, tilted 17° from the orbits of the other planets—and so eccentric it sometimes comes closer to the sun than Neptune. When Pluto was discovered, astronomers initially estimated it to be as large as Earth. They have since learned it is much smaller—only about 1,400 miles wide—smaller even than Earth's moon. Some doubters say Pluto should be classified as a member of the Kuiper belt, a ring of icy bodies beyond Neptune. And even some astronomers defending the planetary status admit that were Pluto discovered today, it might not be awarded planethood. But that's small consolation to planetarium visitors. As one of its docents revealed, a visiting child once asked, "Did you forget my friend Pluto?"

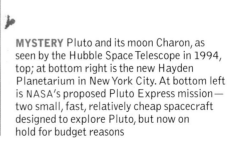

MYSTERY Pluto and its moon Charon, as seen by the Hubble Space Telescope in 1994, top; at bottom right is the new Hayden Planetarium in New York City. At bottom left is NASA's proposed Pluto Express mission—two small, fast, relatively cheap spacecraft designed to explore Pluto, but now on hold for budget reasons

Radiant Message

The Van Allen radiation belts: no human name has ever been given to a more majestic feature of Planet Earth. U.S. scientist James Van Allen's investigation of these cosmic rays—visible from Earth only in the form of the shimmering aurora borealis, or northern lights—rode in tandem with the evolution of the U.S. exploration of space.

Van Allen first sent Geiger counters and other radiation-measuring devices into space after World War II aboard captured German V-2 rockets, then via U.S. Aerobees, one of the first American rockets. Working in the physics department at the State University of Iowa, he sent equipment aloft in cheap plastic balloons, then cobbled together a balloon-rocket device, the "Rockoon." By now an expert on the miniaturization of instruments, he worked with both the Navy's Vanguard and the Army's Jupiter C rockets to send his equipment aloft in the mid-'50s.

When the Soviet Union launched its Sputnik satellite in October 1957, the U.S. satellite program was put on a fast track. The first U.S. satellite, Explorer I, carried Van Allen's equipment into orbit on Jan. 31, 1958. Two months later, Explorer III yielded the data that amazed first Van Allen and then the world: Earth was surrounded by a belt of intense radiation. From Pioneer III, sent aloft late in 1958, Van Allen learned that there are not one but two radiation belts, with a low-intensity slot between them. Van Allen concluded that the outer belt is made of weaker particles, protons and electrons radiating from the sun. At its outer edges, it curves downward in "horns" that hit the atmosphere near Earth's magnetic poles. Van Allen had solved the mystery of the northern lights—a luminous calling card of the sun's interaction with its third planet.

Photograph by Joseph Van Os—The Image Bank

AURORA RISING The northern lights illuminate the skies over Canada's Northwest Territories. The origin and nature of the mysterious lights was unclear until James Van Allen's work on cosmic rays explained them.

VAN ALLEN'S "ROCKOONS"

TIME *profiled James Van Allen on its cover on May 4, 1959. An excerpt:*

At the State University of Iowa, Van Allen was on a slim budget. He launched cheap plastic balloons from the stadium. After V-2s and Aerobees, it was a comedown.

"Wouldn't it be easier," a friend had once asked him, "to lift a rocket on a balloon above most of the atmosphere, and then fire it?" Van Allen decided that the trick should work. He wrangled small, cheap rockets through a friend at the Jet Propulsion Lab; a ballon-rocket combination to carry an 8-lb. payload of instruments 75 miles up was put together for $750.

The Coast Guard agreed to put Van Allen and his Rockoons aboard the icebreaker *Eastwind* bound for Greenland, where cosmic rays are deflected toward the Magnetic Pole by the Earth's magnetic field.

The first ballon rose properly to 70,000 ft., but the rocket did not fire. The second Rockoon behaved in the same maddening way. On the theory that extreme cold at high altitude might have stopped the clockwork supposed to ignite the rockets, Van Allen heated cans of orange juice, snuggled them into the third Rockoon's gondola and wrapped the whole business in insulation. The rocket fired.

NASA: WHAT WE LEARNED ON THE MOON

1. The moon is not a primordial object; it has internal zoning similar to Earth's. It has a thick crust (60 km), a fairly uniform lithosphere (60-1000 km) and a partly liquid asthenosphere (1000-1740 km).

2. The extensive record of meteorite craters on the moon provides a key to the time scales for the geologic evolution of Mercury, Venus and Mars.

3. The youngest moon rocks are virtually as old as the oldest Earth rocks. The earliest evidence of events that probably affected both can be found only on the moon.

4. The moon and Earth are related and formed from different proportions of a common store of materials.

5. The moon is lifeless; it has no living organisms or native organic compounds.

6. Moon rocks originated through high-temperature processes with little or no involvement with water.

7. Early on, the moon was melted and formed a "magma ocean" many tens of kilometers deep.

8. A series of huge asteroid impacts created basins that were later filled by lava flows. Volcanic-fire fountains produced orange and green glass beads.

9. The moon is slightly asymmetrical, perhaps owing to Earth's gravitational pull.

10. The surface of the moon contains a unique radiation history of the sun, which can help explain climate changes on Earth.

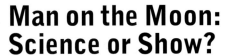
Man on the Moon: Science or Show?

We all know the name of the first man to set foot on the moon—but it's a good guess most of us don't know the name of the moon's most recent visitor. The honor belongs to U.S. astronaut and geologist Harrison Schmitt, of Apollo 17, the final mission of the lunar-landing program. Schmitt, along with crewmate Eugene A. Cernan, discovered orange-colored soil on the lunar surface and left behind a plaque reading, in part, HERE MAN COMPLETED HIS FIRST EXPLORATION OF THE MOON, DECEMBER 1972 A.D.

Almost 30 years have passed since men visited the moon. And over that time, as robot explorers have probed the planets and the Hubble Space Telescope has brought us new visions of deep space, a question festers: Was Neil Armstrong's "giant leap for mankind" a watershed moment in the annals of discovery—or only in the annals of politics? The Apollo program was conceived in the tense days of the cold war, when the high frontier of space became an arena for a propaganda duel. The U.S., inspired by President John F. Kennedy's vow to land on the moon within the decade, put its prestige on the line to achieve a lunar landing.

The Apollo program ultimately placed 12 Americans on the surface of the moon. Of the seven missions that lifted off with hopes of achieving a landing, six were successful, while the Apollo 13 mission was famously plagued by malfunctions; its three astronauts survived only through courage and ingenuity. The sense that the program was more spectacle than science was given a boost when the most prominent images from the Apollo 14 mission were of astronaut Alan Shepard driving a moon buggy around the lunar surface, hitting golf balls.

What did scientists discover from the Apollo program? NASA hears the question so often that its provides a list on its website. An abridged version appears at the left.

Photograph by NASA

MOONWALK Apollo 17 astronaut Harrison (Jack) Schmitt stands on the surface of the moon while holding a rakeful of rock samples in 1972. The lunar buggy *Rover* is at left; it was left behind on the moon

Historic Moments in Space Flight

Mankind's exploration of space has now lasted for the relatively brief span of four decades, yet it has already produced a gallery of great moments, unforgettable people and memorable images—like this one of a certain planet seen from its moon

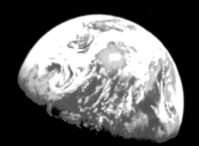

1968

Apollo 8 Orbits the Moon

Only six short years after John Glenn first orbited Earth, Americans achieved two historic firsts: the astronauts of the Apollo 8 mission—William Anders, Frank Borman and James Lovell—became the first men to escape Earth's gravity and the first to orbit the moon. On Christmas Eve, the astronauts sent live TV images from the moon, while reading from *The Book of Genesis*. No doubt the three lunar travelers recalled fellow astronauts Roger B. Chaffee, Virgil (Gus) Grissom and Edward H. White Jr., who had died in a fire on a Cape Kennedy launch pad 11 months before

NASA

1957
Sputnik

The launching of Sputnik, the first Earth-orbiting satellite, by the Soviet Union on Oct. 4, immediately turned America's attention to space. As the New York Times headline at left suggests, the possible use of the satellite as a war machine ("tracked in 4 crossings over U.S.") transformed space from the realm of science (and science fiction) into a new outpost of cold war politics.

1961
Yuri Gagarin: First Man in Space

While U.S. scientists played catch-up in the race to build rockets and launch satellites, the Soviet Union took a giant stride: on April 12, it succeeded in sending cosmonaut Yuri Gagarin on a single-orbit mission around the Earth.

1962
John Glenn Orbits the Earth

With America's political prestige now invested in the space race, the National Aeronautics and Space Administration ramped up its manned-flight Mercury program, naming its first seven astronauts and sending Alan B. Shepard on a suborbital mission on May 5, 1961. Nine months later, John Glenn became the first American to orbit Earth, surviving a potentially lethal problem with his heat shield on re-entry. Glenn later showed off his small Friendship 7 capsule to President John F. Kennedy.

1965
The First U.S. Spacewalk

As the Soviets and Americans continued to play tit-for-tat in space, Soviet cosmonaut Aleksei Leonov became the first man to walk in space outside his Voshkod 2 capsule on March 18. Three months later, on June 3, U.S. astronaut Edward H. White Jr., left, became the first American spacewalker. NASA was now intent on achieving President Kennedy's 1961 goal of reaching the moon by decade's end.

NASA

RALPH MORSE—TIMEPIX (3)

1969

Apollo 11 Lands on the Moon

On July 16, a U.S. Saturn rocket, above, blasted off, carrying three astronauts on their way to the moon. The Apollo 11 team reached lunar orbit three days later, and on July 20, while Michael Collins remained in orbit aboard Columbia, Neil Armstrong and Edwin (Buzz) Aldrin descended to the Sea of Tranquillity in the Eagle module, where Aldrin left his footprint in the lunar dust

■ milestones

1975

The First International Docking

After a decade of dueling, the U.S. and Soviet space programs rode the wave of détente to team up for the first time in space. On July 17, Soviet cosmonaut Aleksei Leonov and U.S. astronaut Thomas Stafford floated through a tunnel linking their craft and shook hands, fellow travelers.

NASA

CHARLES TRAINOR—MIAMI DAILY NEWS/TIMEPIX

NASA

1981

The First Space Shuttle Flight

The U.S. ushered in a new age in the exploration and utilization of space with the launch of a reusable space vehicle, the shuttle. Launched atop a rocket, left, *Columbia* used its stubby wings to land like an airplane. The first flight of the craft, which was plagued by glitches in development, was made on April 12 — 20 years to the day after Yuri Gagarin's orbital flight.

2001

Mir Falls to Earth

The Soviets' orbiting space lab, Mir, was launched in 1986 and was expected to last for perhaps a decade. But year after year, Mir just kept on going, weathering a particularly bad time in 1997, when it survived a collision with an unmanned cargo vessel. Mir finally fell to earth, breaking up into pieces, on March 23, 2001.

1986
Shuttle *Challenger* Explodes on Lift-Off

The pioneering days of America's journey into space were remarkably free of setbacks, given the high degree of danger involved in the enterprise. Though there were plenty of scares, the only tragedy was the launch-pad fire that claimed the lives of three Apollo astronauts early in 1967.

America's good fortune ran out on Jan. 28, 1986, when the space shuttle *Challenger* exploded only 73 sec. after lift-off, killing all six NASA astronauts aboard, as well as amateur astronaut Christa McAuliffe, an ebullient New Hampshire schoolteacher. The tragedy was traced to a crucial set of gaskets in the rocket booster, the O-rings, which failed owing to unseasonably cold temperatures.

CARL SAGAN
1934-1996

1934
Born in Brooklyn, NY

1954
Earns a B.S. from the University of Chicago at age 19. Earns an M.S. and a Ph.D. (all from Chicago) by age 25

1954-60
Develops theories (contrary to current consensus) that Venus was scorchingly hot and that changing colors on Mars were caused by windstorms. Both theories were later confirmed by space probes

1977
Publishes *The Dragons of Eden: Speculations on the Evolution of Human Intelligence*. It wins the Pulitzer Prize

1979
Publishes *Broca's Brain: Reflections on the Romance of Science*, inspired by the pickled brain of Paul Broca, a 19th century anatomist, in a Paris museum

1980
Publishes *Cosmos* and stars in the 13-part PBS series; it wins 3 Emmy Awards

1985
Publishes *Contact*, his only novel

1992
Rejected for membership in the National Academy of Science, perhaps due to his high pop-culture profile

1996
Dies of pneumonia arising from myelodysplasia (a bone marrow disease) in Seattle, Wash.

Wrong End of the Scope?

Astronomers are encouraged to study big stars, not become them. So when Carl Sagan cast himself as a carny barker for the heavens, he caught some flak

SCIENCE HAS A LONG TRADITION OF modesty and understatement, even of calculated obfuscation, so that only an élite priesthood will be privy to its secrets. In these circles, "popularizer" is a pejorative, a term reserved for those who seek the wide angle of the celebrity spotlight. So it was not surprising that fellow scientists sometimes looked askance at Carl Sagan, the ebullient explicator who made stargazing hip, and who seemed as comfortable on Johnny Carson's sofa as he did in an observatory. But Sagan was sui generis: a vital intellect who reconciled the roles of scientist and showman, he found his role as down-to-earth pitchman for the way-up-there.

When he appeared on TIME's cover in 1980, Sagan was 45 years old, a respected Cornell professor and the host of the surprise Public Television hit *Cosmos*, which eventually reached a global audience of 500 million in 60 countries. His books, ranging from speculations about life beyond the earth (*The Cosmic Connection*) to ruminations about the reptilian ancestry of the human brain (the Pulitzer-prizewinning *The Dragons of Eden*), sold millions of copies in a dozen languages.

Son of a U.S.-born mother and a Russian-immigrant father, Sagan loved to gaze at the stars—those he could see from Brooklyn. When he learned in a library book that the stars were enormously distant suns, he told TIME, "this just blew my mind." It remained blown, and the hook worked its way in deeper when young Carl stumbled onto science fiction. A brilliant student, he entered the University of Chicago at 16 on a scholarship; nine years later, he left with a Ph.D. in astronomy and astrophysics.

Sagan was soon making waves in the astronomy world: in 20 years he published some 300 papers, including a prescient argument that Venus' atmosphere of carbon dioxide and water vapor would trap solar heat, creating a "greenhouse effect" that would raise temperatures on the surface—a fact that was soon confirmed by Soviet landers. Along the way, he did stints at Harvard, Stanford and the Smithsonian Astrophysical Observatory in Boston.

In 1968 Sagan accepted Cornell's offer to set up a Laboratory of Planetary Studies; soon he was working with NASA as an adviser and scientific investigator on planetary missions. But he tangled with NASA brass over exploration; Sagan believed robots could do the job better and cheaper than humans. Still, just before NASA sent off its twin Pioneers 10 and 11 to Saturn and Jupiter, he persuaded the space agency to attach plaques identifying the ships' earthly origins on the remote chance they might be intercepted in space. The idea could be seen as the template for his career: a legitimate experiment, it was also a triumph of vision and the questing spirit of science over bureaucratic caution—and it garnered worldwide attention for one of Sagan's pet projects, the search for extraterrestrial life.

In the early 1980s, as the cold war heated up from Afghanistan to Central America, Sagan became a forceful advocate against nuclear war, lecturing and writing often about the "nuclear winter" that might overtake the earth if atomic war were to break out. He died young—at only 62—in 1996. The film version of *Contact*, his story of the search for extraterrestrial life, appeared a year later. Though mushy in spots, it was a rarity for Hollywood: a movie that managed to bring the excitement of science to the screen. Like its author, it proved popular. You got a problem with that?

MARVELOUS MACHINE

Voyager 1 was built to last. The bowl-shaped antenna carried above its 10-sided body was 12 ft. in diameter. At 825 kg (1,820 lbs.), the whole machine weighed less than a compact car. It moved through frictionless space effortlessly from its initial thrust, sometimes affected by the gravitational pull of planets but able to correct its course with blasts from small thruster rockets.

The most complex cluster of equipment was housed forward, near Voyager's two TV cameras, and included an infrared radiometer to measure the heat of planets and spectrometers to analyze the composition of the atmosphere. The magnetometers, for locating and measuring magnetic fields, were carried on the opposite side of the antenna, on a derrick-like boom to keep them free from magnetic distortion. Power to run the equipment was supplied by three cylindrical plutonium generators carried below the spacecraft to keep the radiation from affecting the instruments.

NASA

Saturn Encounter

Topping off a rich decade for America in the exploration of the solar system—including flybys of Mercury, Venus and Jupiter, as well as two landings on Mars—the small Voyager 1 planetary explorer craft paid a visit to Saturn in November 1980, sending back the best images yet of that strange and wondrous world. It was 370 years after the giant planet's marvelous rings were first seen by man, in the low-power spyglass built by pioneering Italian astronomer Galileo Galilei.

Commanded only by its own computers—almost as antiquated as Galileo's spyglass by present standards—Voyager soared past Saturn's mysterious moon Titan, approaching to within 4,000 km (2,500 miles) of its shrouded surface. Gathering ever more speed under the tug of Saturn's gravity, the craft plunged downward toward the outer edge of the planet's rings, swirling bits of cosmic debris. Reaching a peak velocity of 56,600 miles per hour, Voyager skirted within 77,260 miles of the planet's banded cloud tops for its nearest approach to Saturn.

Voyager's flyby produced a trove of new information about the second largest member of the sun's family, a swirling gaseous ball, mostly hydrogen and helium, that could encompass 815 Earths. Before Voyager, Saturn was thought to have only six rings; the craft's sensors identified 1,000 more of them, including several within the long-known Cassini division, the largest gap in the rings. Voyager also detected three small new Saturnian moons, bringing the known total from 12 to 15, and revealed two red spots, some 12,000 km (7,500 miles) in diameter, on Saturn's surface, probably storms not unlike the one responsible for Jupiter's famed red spot.

Photograph by NASA

RINGED WONDER The golden globe of Saturn wears a dark belt of shadows cast by the planet's rings. The two brightest spots are the planet's moons Dione, left, and Tethys

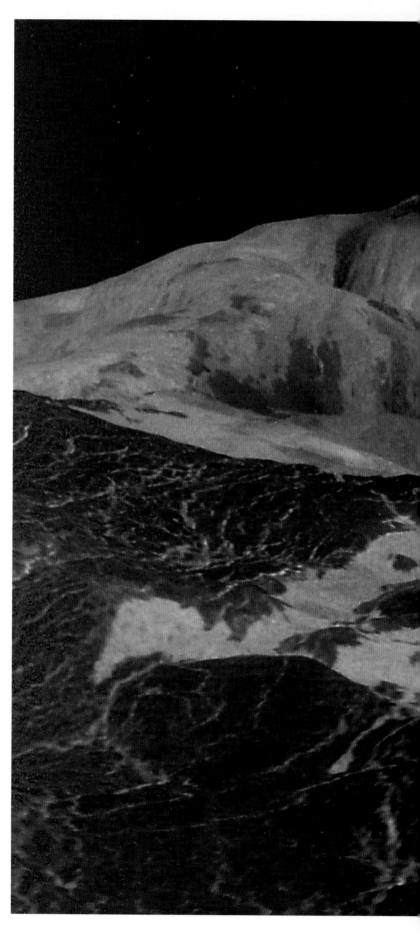

Venusian Volcano

The surface of Venus is far from hospitable. Temperatures hover around 470°C (900°F), the result of a runaway greenhouse effect, and the pressure of its atmosphere is some 90 times that of Earth. Lead would flow like water on Venus, and water cannot have existed in liquid form for perhaps three million years. In 1991 NASA released images take by its Magellan spacecraft that added another element to our vision of the planet: it has been shaped—and perhaps is still being shaped—by massive volcanoes.

Venus is often considered a sister planet to Earth: they are much the same size (Earth is 7,926 miles in diameter, Venus 7,520 miles); they are not far apart in space (Earth orbits the sun at an average of 93 million miles, Venus at 67 million miles); their density is also similar. But Magellan's imaging radar, which employs bursts of microwave energy somewhat like a camera flash to illuminate the planet's surface, showed us just how unearthly a place it is.

Magellan's most stunning image was of Venus' second tallest mountain, Maat Mons, right, which rises five miles above the planet's plain. Its most stunning discovery was that the entire surface of the planet seems to have been reshaped some 500 million years ago—though the planet itself is 4.6 billion years old, the same age as Earth. Scientists based their conclusion on the relatively recent craters found on the surface, where there is no wind erosion. The hypothesis: some event or events 500 million years ago must have resurfaced the planet. The new surface layer may have been the work of massive outpourings of lava from planet-wide volcanic eruptions; huge lava channels can be seen on the surface. Although Venus may still have active volcanoes, no visible fresh lava flows have yet been detected.

Photograph by JPL/NASA

A MOUNTAIN ON THE EVENING STAR
Scientists initially thought the bright-colored patches on Maat Mons were fresh lava flows, but their age remains undetermined

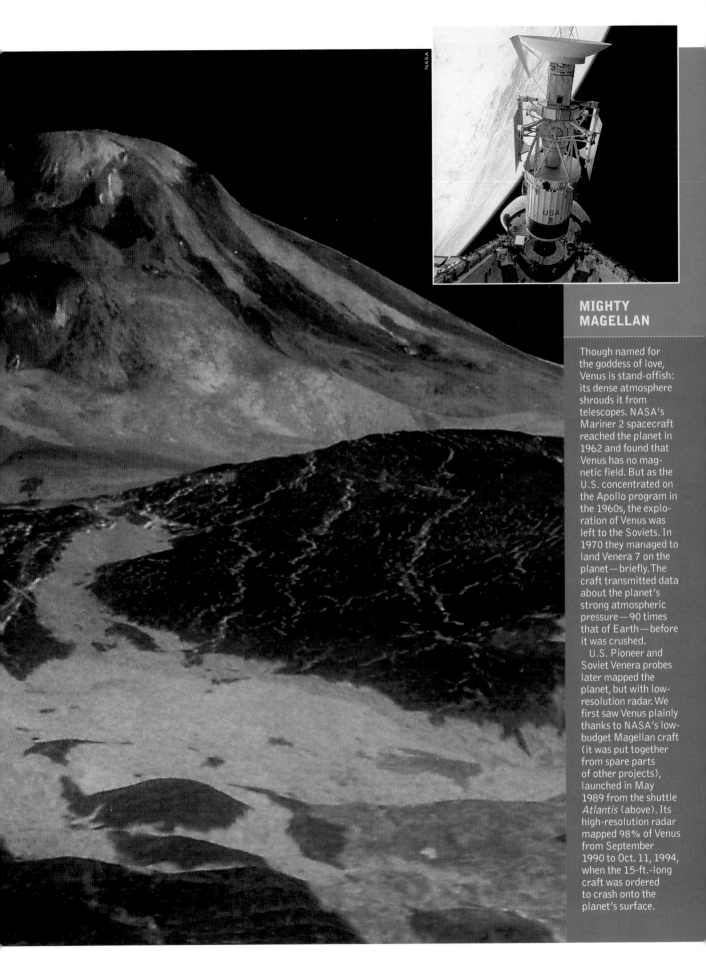

NASA

MIGHTY MAGELLAN

Though named for
the goddess of love,
Venus is stand-offish:
its dense atmosphere
shrouds it from
telescopes. NASA's
Mariner 2 spacecraft
reached the planet in
1962 and found that
Venus has no mag-
netic field. But as the
U.S. concentrated on
the Apollo program in
the 1960s, the explo-
ration of Venus was
left to the Soviets. In
1970 they managed to
land Venera 7 on the
planet—briefly. The
craft transmitted data
about the planet's
strong atmospheric
pressure—90 times
that of Earth—before
it was crushed.
   U.S. Pioneer and
Soviet Venera probes
later mapped the
planet, but with low-
resolution radar. We
first saw Venus plainly
thanks to NASA's low-
budget Magellan craft
(it was put together
from spare parts
of other projects),
launched in May
1989 from the shuttle
Atlantis (above). Its
high-resolution radar
mapped 98% of Venus
from September
1990 to Oct. 11, 1994,
when the 15-ft.-long
craft was ordered
to crash onto the
planet's surface.

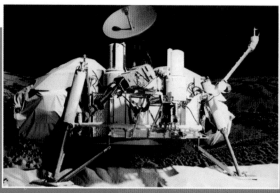

THE VIKINGS
HAVE LANDED

Twenty-one years
before Pathfinder
bounced to a stop on
Mars, the U.S. landed
two craft on the Red
Planet. Vikings 1 and 2
achieved orbit around
Mars in the summer of
1976, but when
photos taken by Viking
1 showed that its
planned landing site
on Chryse Planitia
(the Plains of Chryse)
was crisscrossed by
crevasses and steep
escarpments, its July 4
landing (timed to
coincide with the U.S.
Bicentennial) was
postponed.

Viking 1 finally
touched down on July
20—seven years to
the day after men first
walked on the moon—
on a smooth spot in
the Chryse Planitia
that had been located
by the radio telescope
at Arecibo, Puerto
Rico. Soon it was
sending back the first
pictures of the Mars
landscape, while its
slender robot arm
scooped up samples
of Martian soil and
dumped them into
the $50 million Viking
Lander Biology
Instrument, whose
40,000 components
(packed into 1 cu. ft.)
analyzed their makeup
and age and tested
them for signs of life.
Viking 2 landed a few
weeks later, after
making a major find in
orbit: that the summer
ice cap of Mars' north
pole is made of water.

Two Stars on Mars

There was no one to witness the moment when a visitor from Earth—NASA's 3-ft.-tall Pathfinder spacecraft—landed in the soil of Mars' Ares Vallis floodplain, where the sky is salmon and Earth is a blue morning star. But there was a planet more than 100 million miles away filled with people who were paying heed when it landed, not by coincidence, on July 4, 1997. It had been 21 years since Americans had first sent a visitor to Mars, and this new arrival demonstrated the advance of technology over that period: curled up inside the main Pathfinder lander like a mechanical kangaroo joey was Sojourner, a 1-ft.-tall, 2-ft.-long robot car, or rover, designed to trundle away from the lander and investigate rocks all over the desert-like site.

About that landing: it wasn't exactly soft. Rather, Pathfinder, as planned, hit Mars at 22 m.p.h., rebounded 50 ft. into the thin Martian air, bounced 23 ft. on a second hop and finally rolled to a stop after 92 sec. Cushioned by airbags, Pathfinder absorbed the shock just fine, then unfolded its three petal-shaped solar panels, deflated its airbags and released Sojourner. After a few communication glitches were solved, Sojourner began crawling across the surface at 2 ft. a minute., "driven" by commands sent from Earth.

Though they were scheduled to explore Mars for only 30 days, the two craft remained active for three months. They confirmed, among other findings, the idea that floods of water had poured over the Martian surface in earlier times and discovered that Mars, like Earth, has an inner metallic core. In all, Pathfinder took some 16,000 pictures of the rocky landscape of Ares Vallis, while Sojourner conducted 20 detailed chemical analyses of the soil. Just as their work was done, NASA's Mars Global Surveyor settled into orbit around the planet and began a two-year mission of photographing the topology and atmosphere of Mars with a special camera.

Photograph by NASA

SUPER SKATEBOARD The robot vehicle Sojourner is ready to be deployed off the Pathfinder lander. It was controlled by radio signals sent from Earth

Destination: Mars

If you don't succeed—fly, fly again. So goes the creed at NASA, which followed up the triumphant 1997 landing of the Pathfinder probe on the Red Planet's surface with two hugely embarrassing flops in 1999. In September of that year, the $125 million Mars Climate Orbiter probably burned up as it entered the planet's atmosphere. Reason: engineers who designed the craft confused the American and metric systems of measurement. Only 10 weeks later, the $165 million Mars Polar Lander crashed on the planet's surface; the cause remains unknown.

The probe at right, the Odyssey spacecraft, carries NASA's hopes of redemption. A Delta rocket lifted off April 7, 2001, carrying the geological instrument, scheduled to reach orbit around Mars in October 2001 and remain there for 2½ years, studying minerals in the rocks and measuring for chemical elements like hydrogen in a quest to find water either on or beneath the planet's surface. Burned by its two failures, NASA boosted Odyssey's budget (raising the mission cost to $297 million) and claimed it exhaustively double-checked some 22,000 parameters in the craft's computer software, any one of which could doom the mission if wrong.

"NASA's main goal here is looking for life," Phil Christensen, an Arizona State University geologist, said, "and so life means looking for water." Christensen hopes to find hot springs on Mars: an infrared camera aboard Odyssey will search for them on the planet's dark side. After the mission's launch, a humbled Ed Weiler, head of NASA's space science program, told reporters, "Instead of popping champagne, I popped a Pepsi. When we get all the science down in two years, that will be the champagne." Cheers!

Illustration: NASA

PROBE The orbiter carries three main devices: the Thermal Emission Imaging System will map the mineralogy and morphology of the Martian surface; the Gamma Ray Spectrometer will map the elemental composition of the surface and determine the abundance of hydrogen in the subsurface; the Mars Radiation Environment Experiment will measure aspects of the near space radiation environment to determine any risk to human explorers

VISITING MARS? TAKE GALOSHES

Once upon a time, the surface of Mars was sloshing with oceans and running with rivers. Billions of years ago, however, the low-gravity planet had both its air and water leak away, causing it to become the dead, freeze-dried place it is today. That, in any case, was the prevailing thinking among planetary scientists—until June 2000, when NASA released a flurry of new images from the Mars Global Surveyor spacecraft (which reached Mars orbit in 1997) that suggest that even today, water may be flowing up from the Martian innards and streaming onto its surface—increasing the chance that at least part of the planet is biologically alive.

The tens of thousands of images the Surveyor orbiter has beamed to Earth since are full of hydro-scarring—tracks where vanished rivers once flowed. But a few of the water channels look as fresh as the day they were formed, leading researchers to posit that that day may have been remarkably close to the present one. In the picture above, the dark streaks are believed to be relatively recent water channels.

Ouch!

Scientists call them NEOS—near earth objects—asteroids and comets that cross or come close to Earth's orbit. And sometimes, as the picture of Arizona's Meteor Crater at right shows, near is not the correct term. Were the asteroid that left its scar tissue in the desert some 50,000 years ago to land here today, it might kill millions of people.

Not so long ago, the notion that Earth might be imperiled by falling objects seemed ludicrous, except to Chicken Little. But thanks to the work of the late Eugene Shoemaker, the renowned U.S. Geological Survey astronomer and geologist who was killed in a car crash in 1997 while hunting for craters in Australia's outback, what once seemed like science fiction is now pure science.

In 1973, on his own initiative, Shoemaker began the world's first systematic watch for NEOS. Before his death, he and his partners— first astro-geologist Eleanor Helin, then his wife Carolyn—had discovered more than 800 asteroids, several of them NEOS, and Carolyn herself had discovered 32 new comets. Today astronomers estimate that as many as 2,000 "Earth-crossing" asteroids wider than a kilometer (3,300 ft.) are out there, with potential impact energies ranging from 100,000 to many millions of kilotons. Any of them is capable of causing a worldwide, civilization-threatening catastrophe. Yet only about 7% of them have been located and tracked.

In 1996 two astronomers at a University of Arizona observatory spotted an asteroid perhaps almost one-third of a mile wide just four days before it flew by, missing Earth by some 280,000 miles. Its impact would have rivaled the explosion of every nuclear weapon on the planet—in a single place. Tom Gehrels, another University of Arizona astronomer, thinks that the population of Earth-crossing asteroids larger than 60 ft. across—one of which could destroy London—is as high as 100 million. In Arizona, reason to believe is right out the back door.

Photograph by Jonathan Blair—Woodfin Camp

CALLING CARD The diameter of Metor Crater, right, is 4,000 ft. Scientists believe a 1908 explosion in Tunguska, Siberia, is the last major impact between an asteroid or part of a comet and Earth; it felled trees, started fires and killed reindeer over 850 sq. miles

SMASH-UP ON JUPITER

Eugene Shoemaker helped establish the threat of a comet or asteroid hitting Earth. Fittingly, it was Shoemaker and his associate David Levy who discovered the comet that proved him correct: Comet Shoemaker-Levy 9 split into 21 fragments before hammering Jupiter in 1994

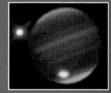

July 18, 1994: A time sequence of Fragment H hitting Jupiter. The blob at center is the impact of Fragments D and G. Top left: Jovian moon Ganymede

Fragment H hits, showing as a tiny red spot at bottom left

Seven minutes later, its fireball is near its maximum brightness

Sixteen minutes later, the fireball has faded and flattened

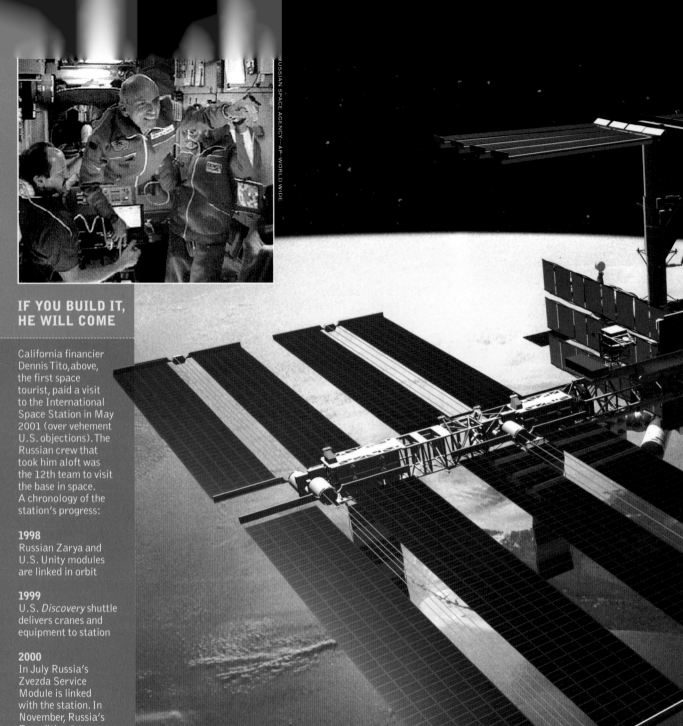

IF YOU BUILD IT, HE WILL COME

California financier Dennis Tito, above, the first space tourist, paid a visit to the International Space Station in May 2001 (over vehement U.S. objections). The Russian crew that took him aloft was the 12th team to visit the base in space. A chronology of the station's progress:

1998
Russian Zarya and U.S. Unity modules are linked in orbit

1999
U.S. *Discovery* shuttle delivers cranes and equipment to station

2000
In July Russia's Zvezda Service Module is linked with the station. In November, Russia's Expedition 1 team begins a four-month stay. In December a five-man U.S. team installs solar arrays

2001
By the end of May, the U.S. Destiny Laboratory Module, the Italian Leonardo Reusable Transport Module and the Canadian Canadarm2 robot arm had been deployed in three separate missions

White Elephant?

On Nov. 20, 1998, the first piece of perhaps the most ambitious multinational engineering project in history—the International Space Station—was launched by a Russian rocket from the Baikonur space center in Kazakhstan. But a question bedeviled its ascent, and still festers three years later: Will the station serve as a scientific platform, as its proponents claim, or is it a colossal white elephant in orbit, a bankrupt idea launched from a bankrupt land? Critics of the manned station contended much of the scientific work proposed to be conducted aboard it could be done just as well, and far more cheaply, aboard unmanned craft.

The notion of putting a permanent manned U.S. platform in orbit was first proposed by Ronald Reagan in his State of the Union address in January 1984. For all the station's great size—it is envisioned to be 296 ft. long and to weigh 500 tons when complete—Reagan viewed it as a fat-free piece of engineering, a lean $8 billion dream machine. What Reagan could imagine and what engineers could build, however, turned out to be two different things. To spread out the expense—and to help keep Russia's space program afloat after the cold war's end—NASA invited international partners to join in. By the time that first component, Russia's Zarya module, which provides the station's propulsion and power, was launched, the project was three or five or 12 times over budget, depending on whose numbers you believed.

Even so, the first launch was followed in short order by the launch of the U.S. Unity module, a six-porthole docking pod. In all, 12 missions to the station had been concluded by June 2001—an appropriate date, give the space station's long odyssey.

Illustration by NASA

UNDER CONSTRUCTION The International Space Station, as NASA projects it will appear when completed in 2006. Solar arrays (wingspan 356 ft.) provide power; an escape vehicle rests beneath a docking module

Wonders of the Solar System

Asteroids with their own moons, ice clouds on Mars, comet rain—as scientists learn more about our small portion of the universe, it seems ever more unpredictable and fascinating

In a Comet's Breakdown, Clues to Life's Start-Up?

Did comets bearing water and organic chemicals help bring life to Earth? The theory got a boost in 2001, when scientists analyzed a Hubble photograph of Comet C/1999 S4 Linear, above, breaking apart as it fell into the sun, shedding mountain-sized chunks of rock and vaporizing tons of ice as it disintegrated. Study of comets Halley, Hale-Bopp and Hyakutake had revealed they had different water chemistry than Earth's oceans. But the characteristics of Linear's breakup suggest its water chemistry matched that of Earth.

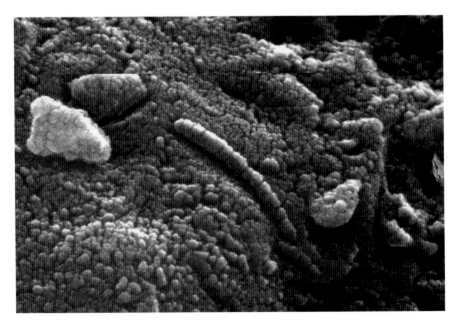

NASA (2)

Fossils in a Rock from Mars?

Meteorite ALH84001 (above, one fragment) is believed to have once been part of Mars. Discovered in Antarctica in 1984, it was found in 1994 to contain tubelike structural forms suggestive of life, left. Scientists remain uncertain as to the nature and origin of the forms.

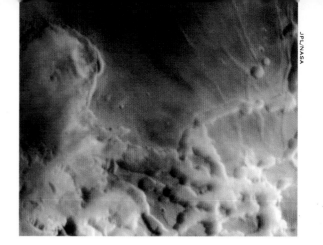

It's Raining Comets!

Hardly anyone took University of Iowa physicist Louis Frank seriously when he first proposed in the 1980s that Earth was being bombarded by cosmic snowballs at the rate of as many as 30 a minute. Every day, he suggested, tens of thousands of icy comets, each the size of a small house and containing 40 tons of water, were vaporizing in the upper atmosphere and raining down on Earth. But in the '90s, pictures from a new NASA satellite, the Polar, left, showed such comets burning up in the atmosphere; the water falls to Earth in the form of rain. That was enough to sell many—but not all— scientists on Franks' theory.

NASA's Polar satellite shot this time-lapse photo of a comet hurtling toward Earth.

Thousands of these cosmic snowballs bombard the planet every day ...

... breaking up before they hit the ground ...

... vaporizing in the upper atmosphere, collecting into clouds and mixing with the rain.

TIME Diagram by Joe Lertola

Ice Clouds on Mars

As the sun rises over Noctis Labyrinthis on Mars, above, bright clouds of water ice can be observed in and around the tributary canyons of the rust-colored desert of this high-plateau region of the planet. The clouds cling to the canyons, suggesting their water condenses in shaded areas in the afternoon, then is vaporized by the morning sun. This image was taken by the Viking probe in 1976.

Asteroid Moons

When NASA's Galileo probe flew by the asteroid belt between Mars and Jupiter in 1993, it came upon an unexpected phenomenon: asteroid Ida, right, had its own satellite, the tiny dot at far right, which was christened Dactyl. Ida is about 35 by 15 by 13 miles in size; Dactyl is less than a mile in diameter. Scientists believe it may have split off from Ida in a collision.

Outer Planet Rings

Everyone knows the glorious rings of Saturn, but as we explore the solar system in greater detail, scientists are finding planetary rings are common. At left is a false-color view of the nine rings of Uranus, taken by Voyager 2 in 1986. Voyager 1 discovered Jupiter's ring (or rings) in 1979; the current argument is whether the phenomenon is one ring divided into four sections—a main ring, an inner "halo" ring and two outer "gossamer" rings—or four separate rings.

EDWIN
HUBBLE
1889-1953

1889
Born in the
small town of
Marshfield, Mo.

1910
Graduates with
a B.A. from the
University of
Chicago, then
studies as a Rhodes
scholar at Oxford

1919
Begins working
at Mount Wilson
in California, after
serving in France
in World War I

1923
Proves that the
universe extends
beyond the edges
of the Milky Way
galaxy

1925
Creates the first
useful scheme for
classifying galaxies

1929
Proves that the
universe is expanding

1931
Albert Einstein
visits Mount Wilson
to thank Hubble,
whose theory of
the expanding
universe validated
Einstein's work on
general relativity

1936
Publishes *The
Realm of the
Nebulae,* a popular
account of his work
and theories

1943
During World
War II, he is the
temporary head of
ballistics at the
Aberdeen Proving
Ground in Maryland

1953
Dies on Sept. 28

A Lever To Move the Stars

Archimedes claimed he could move the world, given a long enough lever. Using a telescope as his tool, Edwin Hubble moved both our world—and the cosmos

AS THE 20TH CENTURY DAWNED, MOST astronomers believed the Milky Way galaxy, a swirling collection of stars a few hundred thousand light-years across, made up the entire cosmos. But peering deep into the reaches of space, Edwin Hubble discovered that the Milky Way is not unique; it is just one of millions of galaxies that dot an incomparably larger setting. And he went on show that this galaxy-studded cosmos is growing larger, inflating majestically like an unimaginably gigantic balloon. Hubble did nothing less, in short, than invent the idea of the universe and then provide the first evidence that led to the Big Bang theory, which describes its birth and evolution.

Hubble was born in Marshfield, Mo., in 1889. Always interested in astronomy, he graduated from the University of Chicago, then took a Rhodes scholarship at Oxford and studied law for one (very boring) year. Then he went to the Yerkes Observatory in Wisconsin to make friends with the sky. After serving in France in World War I, he took a long-promised job at the observatory at Mount Wilson in California.

Even as an apprentice astronomer, Hubble concentrated on the nebulae—the faint patches of light scattered among the stars. His first step at Mount Wilson was to find out how far away they were. His general method was to determine the intrinsic brightness of objects in a nebula and then gauge its distance (the fainter it was, the farther away it was). Variable stars called Cepheids told him that the bright nebula called Messier 31 was 680,000 light-years away. Messier 31 was therefore no mere part of the Milky Way galaxy but an isolated star system far out in space, and as big as our entire galaxy.

Other, longer-range measuring sticks carried him farther on his march into space. The most distant nebulae showed as tiny, dim blobs. By a complex statistical method, Hubble proved after years of work that these dimmest glimmers were so far away that their light took 500 million years to reach Mount Wilson. This accomplishment led to his crowning discovery—the theory of the expanding universe.

Astronomers have a speedometer to clock the motions of skittish heavenly bodies. They take spectrographs: photographs of the body's light spread out by a prism into a band of colors. If the band is "shifted toward the red" (i.e., if it is redder than normal), it shows that the body is moving away from the earth. Studying the light of the distant nebulae, Hubble found in every case a red shift. The farther off a nebula was, the faster it appeared to be rushing away, and the enormous speeds (thousands of miles per second) were unexpected, strange and startling.

Hubble reached a momentous conclusion: that the speed of recession of the nebulae is directly proportional to their distance. This means that each of the large units of matter in the universe is moving away from every other unit. The Milky Way nebula is not the center of explosion; every nebula is an explosion center. Casting around for a layman's analogy, Hubble compared the expanding universe to a rubber balloon with small dots (representing nebulae) spaced equally far apart on its surface. When the balloon is blown up larger, each dot becomes more distant from every other dot. Place an observer on any dot, and every other dot-nebula will be moving away from him.

In science, progress often takes small steps. But Hubble opened our eyes to a vast, dynamic cosmos, far bigger, far older and far more complex than we had ever dreamed.

A HUBBLE PRIMER

LAUNCHED
April 24, 1990

REPAIRED
December 1993.
Missions to install
new equipment took
place in 1997 and
1999. Next planned
update: 2002

WEIGHT
25,000 lbs.
(11,600 kg)

ORBIT
372 miles
(600 km) above
Earth. Speed:
17,000 m.p.h.
A complete orbit
takes about
96 min.

POWER SUPPLY
Two 8-ft. by 40-ft.
solar panels. Nickel
batteries store
energy for use
when the telescope is
in Earth's shadow

CYBERSPACE
NASA maintains a
number of websites
that display Hubble
images and update
news of its findings.
Begin at:
www.nasa.gov

The Orbiting Eye

For years before its launch, the Hubble Space Telescope was promoted by NASA and the astronomers who looked forward to using it as stargazing's Next Big Thing. The huge scope's orbital perch—high above the distorting effects of Earth's atmosphere—would yield high-resolution images of the heavens, we were told. So, more than a disappointment, it was an embarrassment for the U.S. space agency when the Hubble, launched into orbit in April 1990, turned out to have faulty vision: its eyesight was marred by an improperly manufactured mirror.

America had invested too much in the Hubble to allow it to fail. NASA launched a rescue mission to the "eye in the sky" in December 1993, charging the seven astronauts aboard the *Endeavour* with the toughest assignment ever handed to a shuttle crew and the most complicated mission since the moon shots of two decades earlier. Their mandate was not only to sharpen the telescope's flawed vision but also to revamp some faulty electronic systems, put in new gyroscopes and replace two unstable solar-energy panels. Getting the job done would require wrestling huge pieces of machinery into tight spaces, disconnecting and reconfiguring fragile electronic equipment and making sure no loose screws damaged the delicate telescope—all while wearing puffy pressure suits and bulky gloves in a vacuum at zero gravity and at –300°F.

Result: "Piece of cake!" boasted happy astronaut Kathryn Thornton, perched atop the shuttle's 50-ft. robot arm as she sent a mangled solar-energy panel off into space. The Hubble's image was cleared; so was NASA's. Today, the Hubble is making good on its hype. As the pictures on the following pages show, the visions of the cosmos it has yielded are not only valuable to scientists but also awe-inspiring to laymen.

Photograph by **NASA**

EYE EXAM Perched on the robot arm of space shuttle *Endeavour,* astronauts F. Story Musgrave and Jeffrey Hoffman (one partially obscured) work on the Hubble Space Telescope in December 1993

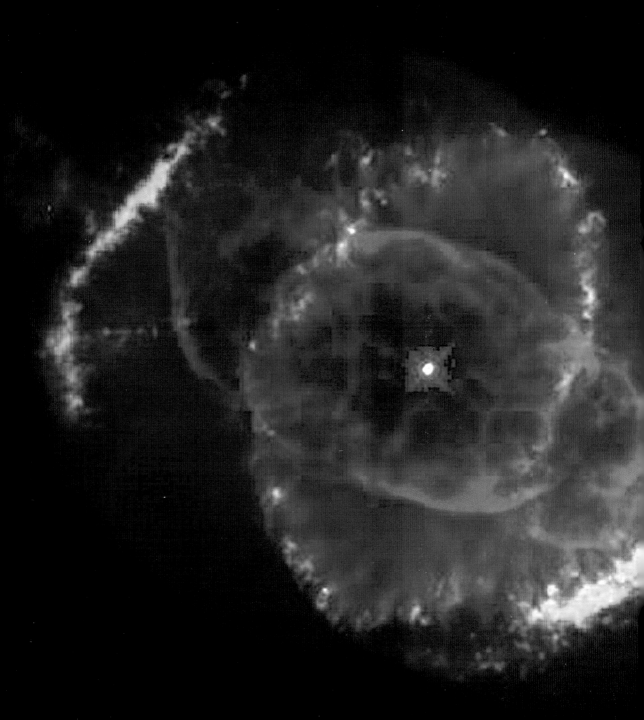

The Discovery Machine

Saturn's auroras. Starmaking factories. Galaxies in collision. The Hubble Space Telescope has given us images of interstellar events that might once have been only an astronomer's wild surmise. And the discoveries continue: in the spring of 2001, a 1997 image taken by the Hubble was declared to confirm the existence of "negative gravity," Albert Einstein's controversial theory about the interaction of space, matter and energy

Dying Double Star

This object, called NGC 6543,
consists of concentric gas shells,
jets of high-speed gas and
unusual shock-induced knots of
gas. Scientists posit that it
consists of the remains of two
dead stars.

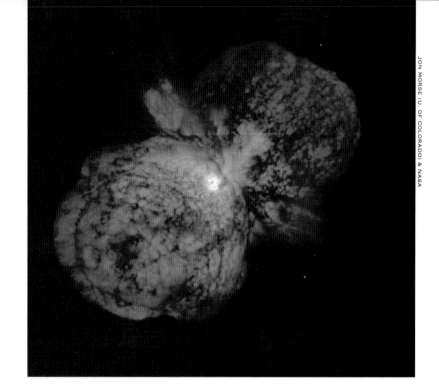

Bipolar Exploding Star

Eta Carinae became one of the brightest stars in the southern sky about 150 years
ago. Its explosion produced two polar lobes and a large thin equatorial disk, all
moving outward at about 1.5 million miles per hour—yet the star itself seems to
have survived. Eta Carinae is estimated to be only 100 times more massive than
the sun, yet it radiates 5 million times the sun's power—we don't know why.

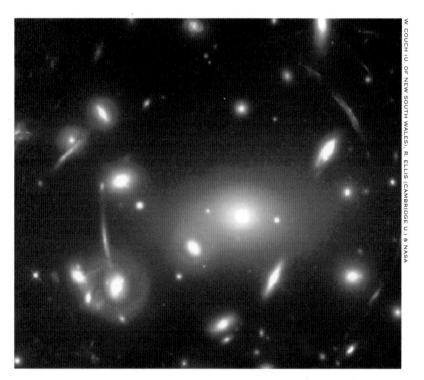

Gravitational Lenses

The Hubble confirmed a main tenet of Einstein's general theory of relativity—
that gravity bends light—in this picture of a rich galactic cluster, Abell 2218.
Many lenses are at work here, magnifying galaxies some five to 10 times further
away than that of Abell 2218. Gravity also distorts and mulitplies the images of
the more distant galaxies. Some distortions appear as arcs between the galaxies.

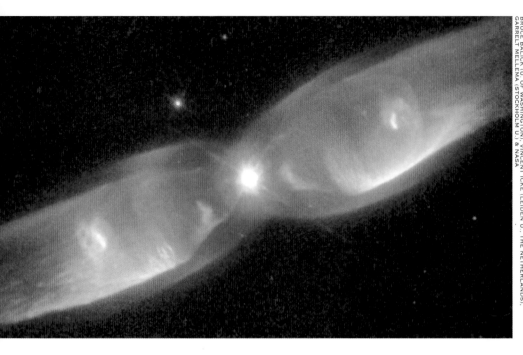

Butterfly Nebula

M2-9, a bipolar planetary nebula 2,100 light-years away, has been dubbed the "Twin Jet" Nebula and a "Butterfly" Nebula. The central star seen here is known to be one of a pair of stars that orbit each other very closely. Some scientists even suspect that one star is being engulfed by the other. Astronomers theorize that the shape of the system is determined by processes similar to those of a jet engine: gravity from one star pulls gas from the surface of the other and flings it into a thin, dense disk; when high-speed wind rams into the disk, it acts as a nozzle, funneling gas into a sort of stellar exhaust trail.

Nebula Bubble in a Star Factory

Intense radiation from newborn, ultra-bright stars has blown a glowing spherical bubble in the nebula N83B (or NGC 1748), below. This star-forming factory is located in a small section of a neighboring galaxy, the Large Magellanic Cloud. The young stars are just emerging from the shelter of their prenatal molecular cloud. The bubble is approximately 25 light-years in diameter.

Dust Disk Around a Black Hole

The image at right shows a spiral-shaped disk of dust fueling a massive black hole in the center of galaxy NGC 4261, 100 million light-years away. By measuring the speed of the gas swirling around the black hole, scientists calculate that the object at the center of the disk is 1.2 billion times the mass of our sun. Yet it is concentrated into a region of space not much larger than our solar system.

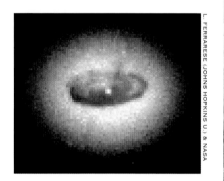

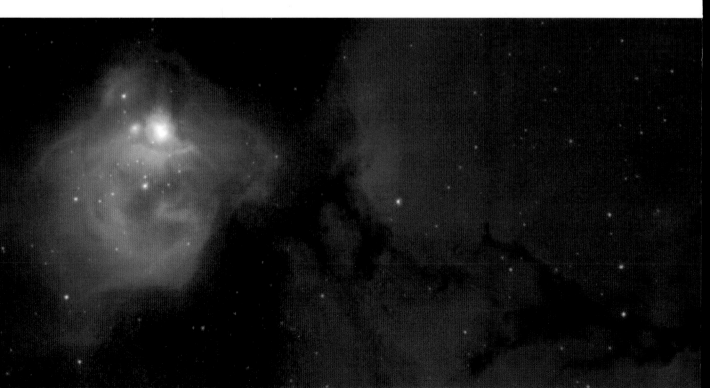

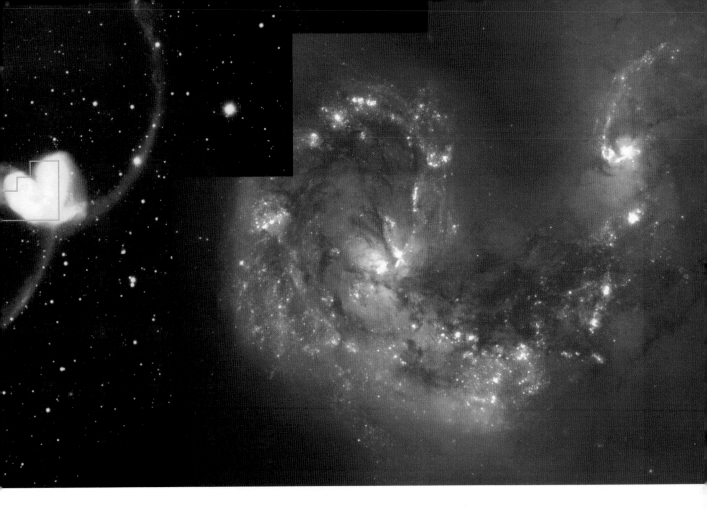

Colliding Galaxies

Hit the brakes—too late! Above are two views of a cosmic collision. At left is a ground-based view of the Antennae galaxies—so called because the long tails of luminous dust resemble an insect's antennae. At right is a Hubble view of a part of the same galaxies. The spiral-like patterns, traced by bright-blue star clusters, reflect a firestorm of star birth triggered by the collision.

Negative Gravity

After studying a 1997 Hubble picture of a distant exploding star, astronomers declared in the spring of 2001 that the image confirmed one of Einstein's conjectures about the universe: that all of space is bubbling with "negative gravity," or "dark energy," which creates a mutual repulsion between objects normally attracted to each other by gravity. The star, seen in three views here, exploded about 11 billion years ago; negative gravity, less powerful then than now, makes it twice as bright as it would appear if Einstein's theory were false.

Where Stars Are Born

This photomontage of pillars of hydrogen gas and dust is perhaps the most familiar image yielded by Hubble—so far. What we don't see here is a flood of ultraviolet light from hot, massive newborn stars (off the top edge of the photo) that is eroding much of the matter around the columns. This formation is part of the Eagle Nebula, or M16, the 16th object in Charles Messier's 18th century catalog of "fuzzy" celestial objects. It is 7,000 light-years from Earth.

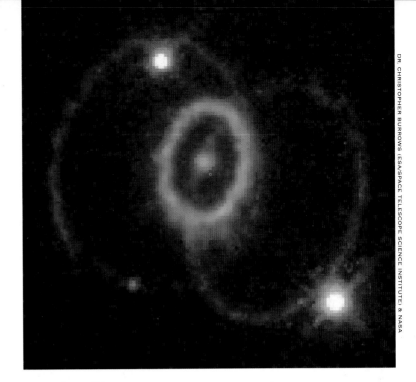

When Stars Die

When the giant star Supernova 1987A exploded, it sent out a debris trail in the form of three rings; the outer two are extremely thin. This is the Hubble's view of the event witnessed in 1987 and featured in the story on pages 100-101. The rings appear to overlap but are on three separate planes. The matter in the rings will expand into space for millions of years. Another name for it: stardust.

The Auroras of Saturn

Northern lights on Saturn? Of course—and southern too. Hubble's Imaging Spectrograph provided this image of Saturn's ultraviolet auroras in October 1997. The auroras were discovered by the NASA probe Pioneer 11 in 1979; Voyagers 1 and 2 (*see pages 76-77*) first mapped them in the early 1980s.

Beyond Hubble

When it comes to taking mind-boggling pictures of deep space, the Hubble Space Telescope stands (well, orbits) alone. But a new generation of telescopes here on Earth, like the twin Keck telescopes on Mauna Kea in Hawaii, is making major contributions to astronomy. And the next wave of scopes, now on the drawing board, will provide a giant stride forward in the study of the skies.

Older telescopes were powered by a mirror that is essentially a huge hockey puck of glass ground into a concave, light-focusing curve on one face and coated with reflective metal. To keep from sagging under its own weight and distorting the curve, the mirror had to be thick: the one at California's Hale Observatory, built in 1949, is a bulky 26 in. thick and weighs 20 tons. That enormous heft called for an even more massive support structure to hold the whole thing up. Scaling the design up any further would have been absurdly expensive.

University of Arizona astronomer Roger Angel's solution to the problem of sagging glass was to cast huge mirrors that are mostly hollow, with a honeycomb-like structure inside to guarantee stiffness. University of California at Santa Cruz astronomer Jerry Nelson opted instead to create a mirror not from a single huge slab of glass but from 36 smaller sheets that would, under a computer's control, act as one.

In Europe, design teams came up with yet a third idea, the exact opposite of Angel's: instead of making the mirror hollow to save weight, let it be thin—about 8 in. thick for an 8-m mirror, in contrast to the 5-m Hale's 26 in.—and counteract the resulting floppiness with computer-controlled supports that continually readjust its shape. With both enormous size and smooth performance, these giant telescopes are doing science on a heroic scale—and with the development of adaptive optics (*see sidebar*) their contribution will only increase.

BIG EYES To make huge mirrors manageable, Roger Angel, top right, casts slabs of glass that are mostly hollow. Bottom left, the Gemini North observatory is located on Mauna Kea, Hawaii. The Gemini mirror, below right, employs relatively thin mirrored sheets whose shape is maintained by computer control

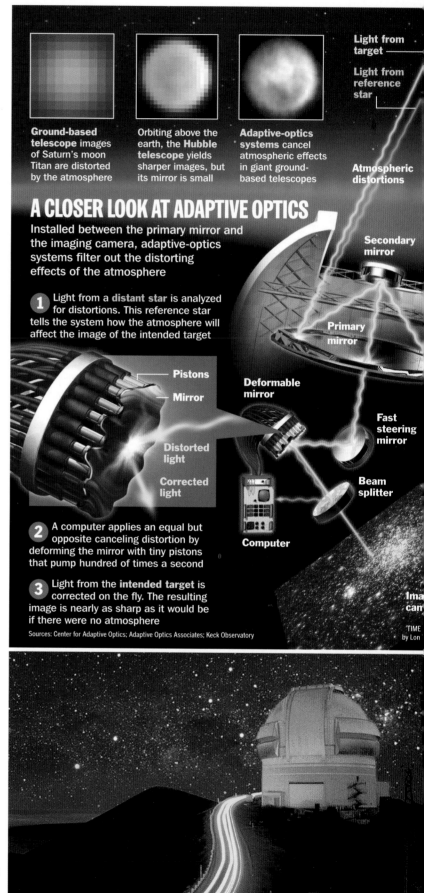

Ground-based **telescope** images of Saturn's moon Titan are distorted by the atmosphere

Orbiting above the earth, the **Hubble telescope** yields sharper images, but its mirror is small

Adaptive-optics systems cancel atmospheric effects in giant ground-based telescopes

Light from target

Light from reference star

Atmospheric distortions

A CLOSER LOOK AT ADAPTIVE OPTICS

Installed between the primary mirror and the imaging camera, adaptive-optics systems filter out the distorting effects of the atmosphere

1 Light from a **distant star** is analyzed for distortions. This reference star tells the system how the atmosphere will affect the image of the intended target

2 A computer applies an equal but opposite canceling distortion by deforming the mirror with tiny pistons that pump hundred of times a second

3 Light from the **intended target** is corrected on the fly. The resulting image is nearly as sharp as it would be if there were no atmosphere

Secondary mirror

Primary mirror

Fast steering mirror

Beam splitter

Deformable mirror

Pistons

Mirror

Distorted light

Corrected light

Computer

Ima can

Sources: Center for Adaptive Optics; Adaptive Optics Associates; Keck Observatory

TIME by Lon

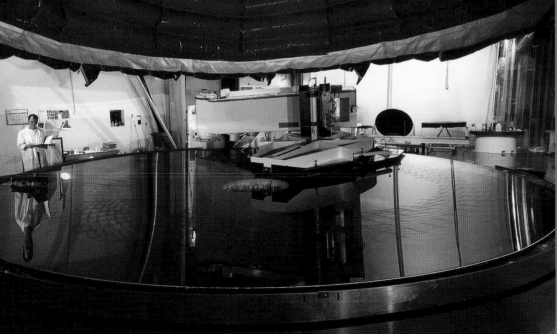

TWINKLE, TWINKLE—NOT!

The Hubble Space Telescope's nearly 2.5-m (8-ft.) mirror isn't all that powerful, but since it floats above Earth's constantly roiling atmosphere, it has been unrivaled in the sharpness of its images. No more. Using an ingenious technological trick to eliminate atmospheric blur, many new telescopes will soon achieve Hubble-quality focus—and even beat it.

The secret: taking the twinkle out of starlight. The means to do this—adaptive optics—was originally developed in secrecy by the Department of Defense to help military snoops take sharp pictures of Soviet spy satellites. Largely declassified in the 1980s, the technology is now being adapted for big telescopes everywhere. The idea is straightforward: stars and galaxies twinkle and shimmer because turbulent pockets of air act as weak, light-distorting lenses (heat rising from a car's hood or an asphalt parking lot causes a similar effect). With adaptive optics, though, a computer can measure the shimmer and cancel it out.

Will adaptive optics make space scopes obsolete? Not entirely. Space is still the best place to take super-sharp pictures in ordinary light. That's why NASA's plan to launch a Next Generation Space Telescope by 2009 still makes sense.

SPACE **99**

Bang!

Feb. 24, 1987: On an 8,000-ft. mountaintop in northern Chile, light that had been traveling through space for 170,000 years passed through the lens of a 10-in. telescope at Las Campanas Observatory and was reflected into a camera set up by Ian Shelton, 29, a Canadian astronomer. For hours Shelton had been taking long exposures of the Large Magellanic Cloud, a nearby galaxy. "It was time to go to bed," he later recalled. But before turning in, he made up his mind to develop the last photographic plate. Lifting it from the developing tank, he scrutinized it, then stopped short. There, near a feature within the LMC known as the Tarantula nebula, was an unfamiliar bright spot.

"I was sure that there was some plate flaw on it," Sheldon said, "but it was no flaw." He walked outside, looked up at the Large Magellanic Cloud and, without a telescope or binoculars, clearly saw an exploding star, or supernova. While hundreds of supernovas occurring in incredibly distant galaxies had already been spotted by powerful telescopes, this was the first one visible to the naked eye since 1885. More important, at a distance of only 170,000 light-years, it was the brightest supernova to appear in terrestrial skies since 1604. News of Shelton's discovery, promptly named 1987A (for the first supernova of the year), was telegraphed to observatories around the world. "It's like Christmas," said astronomer Stan Woosley of the University of California at Santa Cruz. "We've been waiting for this for 383 years."

Photographs by David Malin—
Anglo-Australian Observatory

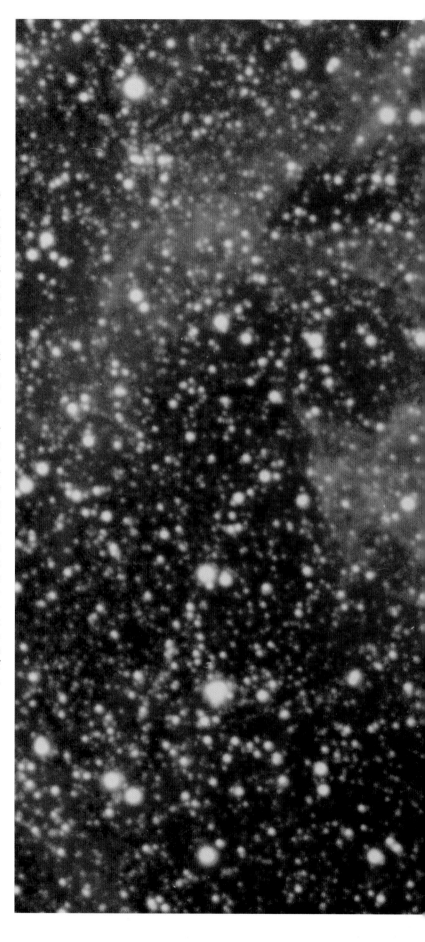

LUCKY STAR Scientists thrilled to witness a supernova in their lifetime: 1987A was the brightest example to appear since pioneer astronomer Johannes Kepler saw one in the constellation Ophiuchus in 1604—five years before Galileo perfected his early telescope

BEFORE

AFTER

INSIDE A SUPERNOVA

The enormous processes that occur during a supernova event produce most of the elements in the universe. These elements are hurled into the cosmos by the force of the supernova blast to form great clouds of gas and dust. Subsequent supernovas send shock waves through the clouds, coalescing gas and dust and starting the formation of new stars and planets. Thus the planets and any life that evolves on them consist of elements forged in supernovas. These stellar explosions also generate energetic particles (cosmic rays) that can cause mutations in terrestrial organisms and may have played a direct role in the evolution of life on earth.

Early readings showed that the shell of gases expanding around 1987A was initially traveling outward at nearly 10,000 miles per second. After several weeks, the supernova changed color, moving from blue to red much faster than scientists had expected, indicating that the rapid expansion of the gas shell was causing it to cool. For a recent view of Supernova 1987A taken by the Hubble Space Telescope, see page 97.

ARECIBO'S HARVEST

The radio telescope in Puerto Rico has contributed to a number of major finds. Among them:

SATURN
Arecibo's radar indicated that the rings of Saturn are composed of chunks of ice and are many layers thick. This was later confirmed by the Voyager spacecraft

VENUS
Arecibo found that the planet Venus rotates clockwise, unlike all the other planets except Uranus

MILLISECOND PULSARS
The radio telescope discovered these neutron stars that have collapsed into dense balls and act as rotating radio lighthouses

COMETS
In 1980 signals from Arecibo bounced off a comet, proving that cometary nuclei are solid, not gaseous

PLANETS IN DEEP SPACE
While working here in 1992, Aleksandr Wolszczan found the first strong evidence of planets outside the solar system

GRAVITY WAVES
Working with pulsars found by Arecibo, Joseph H. Taylor Jr. and Russell Hulse confirmed the existence of gravity waves, as predicted by Einstein. They shared the Nobel Prize in 1993 for their work

Earth's Eardrum

It's commonplace to envision telescopes as stupendous eyes: after all, the term means "far sight." But one of Earth's most productive tools for discovery—the radio telescope embedded in the mountains near Arecibo in Puerto Rico—is not a giant eye, but a giant ear. Conceived in 1958 by electrical engineer William E. Gordon, it was built in 1960 by the National Science Foundation and overhauled in 1997. The giant receiving dish spans 1,000 ft. in diameter, covers 20 acres and is composed of nearly 40,000 perforated aluminum mesh panels, each measuring 3 ft. by 6 ft. A 600-ton platform, where the real work gets done, is suspended 426 ft. over the dish by cables strung from three reinforced concrete towers. The structure is so vast it can be seen by passengers on jets flying overhead at 30,000 ft., yet its entrance road is almost impossible to find on the ground, to discourage unwanted visitors.

There is another major difference between Arecibo and optical scopes: the big dish not only receives information, but it broadcasts as well, bouncing radar signals off objects in the solar system to help scientists glean information from the echoes it receives back. In 1974 the Arecibo Interstellar Message was beamed to M13, a globular star cluster 25,000 light-years away. That signal—essentially saying "We're here" to anyone out there—contained 1,679 bits of information on everything from the chemical makeup of our atmosphere to Earth's position in the solar system. In 1992 Arecibo became part of NASA's controversial SETI (Search for Extraterrestrial Intelligence) program, systematically scanning the heavens for any similar radio transmissions broadcast by other life forms to us. Congress pulled the plug on funding a year later, but the program continues under private sponsorship. Had it produced any great discoveries—well, we suspect you'd have heard the news by now.

Photograph by David Parker—SPL/Photo Researchers

STAR SEARCH Look familiar? The Arecibo radio telescope is a star in its own right: it was featured in the 1995 James Bond flick *Goldeneye* and in the 1997 film *Contact*

companion object, which is thought to be a planet. This may be the first picture ever taken of an extrasolar planet —its nature remains unclear.

Wonders of Deep Space

As new technology helps us probe deeper into the universe, even the images that seem routine may contain the most exciting new visions. In this view, for instance, the small dot on the right may be the first picture we have ever seen of a planet orbiting another sun

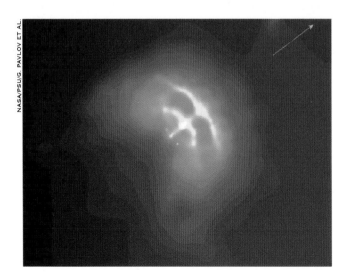

Pulsar Crossbow

Identified in 1967 by Jocelyn Bell and Anthony Hewish, pulsars emit radio waves in regular bursts. The Vela pulsar, the collapsed stellar core of a supernova, left, spins 10 times a second as it hurtles through a debris cloud, emanating jets of high-energy particles in a crossbow shape.

Black Hole

The two images at right are evidence of a massive black hole in galaxy M87. Near right, a long view of the galaxy; far right, a Hubble close-up of the massive disk of gas orbiting the black hole at its center.

Red Giant

Betelgeuse, or Alpha Orionis, is one of the most familiar stars; it marks the left shoulder of the constellation Orion, right. Betelgeuse is a red giant, a star past its prime that has (relatively) cooled and distended to enormous size. If Betelgeuse replaced our sun, its atmosphere would extend beyond Jupiter's orbit.

Quasar

Quasars, discovered by Allan Sandage and Thomas Matthews in 1961, are small objects of ultrahigh luminosity that emit strong radio waves. At left, a single quasar is multiplied by a gravitational lens.

Seven White Dwarfs

A 1995 Hubble telescope image of globular cluster M4, below, shows seven white dwarf stars (circled) among the cluster's brighter population of yellow sunlike stars and red giants. White dwarfs are remnants of older stars that have expended their fuel and subsided into a quiescent stage.

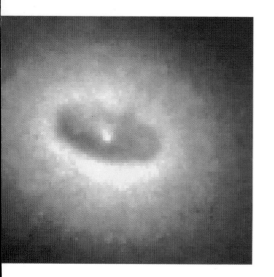

In the Realm of Singularities

First Stephen Hawking probed the secrets of black holes. Then, in his popular books, he managed to help infinitely dense readers like us understand them

THE PLIGHT OF BRITISH ASTROPHYSI- cist Stephen Hawking can be read as a metaphor for the scientific process: bound by maddening restrictions, the inquisitive human spirit finds ways to explore the outer reaches of reality. While ALS—amyotrophic lateral sclerosis, or Lou Gehrig's disease—has made Hawking a virtual prisoner in his own body, it has left his courage and humor intact, his intellect free to roam. And roam it does, from the infinitesimal to the infinite, from the subatomic realm to the far reaches of the universe.

Stephen Hawking does his thinking at Cambridge University, where he is Lucasian professor of mathematics, a seat once occupied by Isaac Newton. His parents were Oxford alums; young Stephen, enthralled by physics, enrolled at Oxford's University College—where he did not distinguish himself. When he moved to Cambridge for graduate work in relativity, he began having physical problems, and doctors soon delivered the bad news: he had ALS, a progressive deterioration of the central nervous system that often brings a fast death. Hawking's illness progressed slowly. But it concentrated his mind: the afflicted graduate student turned with new vigor to his studies.

What particularly intrigued him was singularities, strange celestial beasts predicted by Albert Einstein's theory of general relativity. Einstein's equations indicated that when a star several times larger than the sun exhausts its nuclear fuel and collapses, its matter crushes together at its center with such force that it forms a singularity, an infinitely dense point with no dimensions and irresistible gravity. A voluminous region surrounding the singularity becomes a "black hole" from which—because of

that immense gravity—nothing, not even light, can escape.

Scientists had found compelling evidence that black holes exist, but they were uncomfortable with singularities, because all scientific laws break down at these extreme points. Most physicists believed that in the real universe the object at the heart of a black hole would be small (but not dimensionless) and extremely dense (but not infinitely so). Enter Hawking. He and mathematician Roger Penrose developed new techniques proving mathematically that if general relativity is correct down to the smallest scale, singularities must exist. Hawking went on to demonstrate—again, if general relativity is correct—that the entire universe must have sprung from a singularity.

He later discerned several new characteristics of black holes and demonstrated that the stupendous forces of the Big Bang would have created mini–black holes, each with a mass about that of a terrestrial mountain but no larger than the subatomic proton. Then, applying the quantum theory, Hawking was startled to find that the mini–black holes must emit particles and radiation. More remarkable, the little holes would gradually evaporate and, 10 billion years or so after their creation, explode with the energy of millions of H-bombs. Other physicists, long wedded to the notion that nothing can escape from a black hole, generally came to accept that discovery. And the stuff emitted from little black holes (and big ones too, but far more slowly) is now called Hawking radiation.

In 1988 Hawking managed to elucidate his highly abstruse theories for general readers in his *A Brief History of Time.* The book made him a best-selling author—and for an astrophysicist, that's not only a singularity, it's a Big Bang.

Aliens and UFOs

"The truth," we are told, "is out there." Problem is, we're looking for proof right here. Multiple eye-witnesses claim to have seen humanoid corpses at the site of a 1947 flying saucer crash in New Mexico? The UFO turns out to have been a secret weather balloon sent aloft to spy on the Soviet atom bomb program, and the corpses were dummies used in high-altitude parachute tests—or so says the government. If you want to believe—believe!

Spaced Out

Flying saucers! Faces on Mars! "I want to believe!" Science fiction is a fine form, but in pop culture, the fiction often outpaces the science

CHIP SIMONS

TED SOQUI—COR

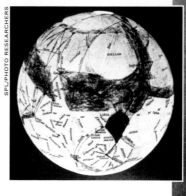

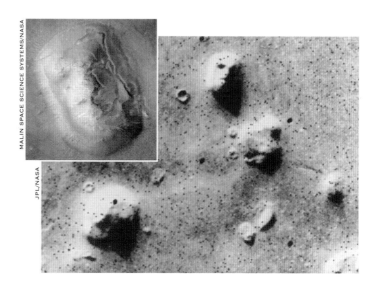

Canals on Mars

Talk about getting lost in the translation: in the 1890s, astronomer Giovanni Schiaparelli used the Italian word *canali* (meaning channels, or grooves) to describe his observations of Mars. But *canali* was translated for U.S. readers as "canals," which inspired American astronomer Percival Lowell, above, to document what he assumed were artificial waterways on Mars, inset. Lowell concluded that the elaborate network of *canali* was built by an intelligent race. It wasn't until July 14, 1965—when the Mariner 4 spacecraft returned the first detailed images of the Martian surface—that Lowell's last holdouts were finally shamed into silence.

Astrology

Since the first-known prediction based on month of birth was carved on a Babylonian tablet sometime around 1200 B.C., the notion that our fates are entwined with the stars has exerted a peculiar hold over the minds of men. The ancients imagined they saw the shapes of animals in the 12 constellations that lie in the sun's path across the sky; hence the zodiac (from Greek for "circle of animals") and the 12 signs that supposedly speak volumes about us. Astrology, last widely popular during the Renaissance (above, a 1531 French woodcut), enjoys its greatest currency during times of learning and discovery. Then again, the Renaissance was also the last time that Neptune, Uranus and Pluto were aligned as they are now.

The Face on Mars

In the summer of 1976, the Viking 1 orbiter, scanning Mars, recorded what seemed to be a giant statue of a human head, above. Scientists saw a trick of the light; conspiracy buffs weren't convinced. So NASA had the Mars Global Surveyor satellite's stronger camera take new photos in '01, inset. Verdict: sometimes a pile or rocks is just a pile of rocks.

Nazca Lines

In the early 1970s, author Erich von Daniken claimed the giant glyphs carved into Peru's Nazca landscape were runway markers for aliens who landed on Earth in ancient times. Scientists believe the lines trace the paths of underground aquifers—or were meant as messages to the gods.

FLASHPOINT
Lava showers from an eruption
on the south east crater of Mt.
Etna in Sicily, 1998

Photograph by Dr. Juerg Alean — SPL/Photo Researchers

Discoveries on
Earth

"I used to dream of flying—the classic attempt
to get away from the reality of Earth. But
since I have been diving, I have not had the
dream. I am miserable out of water. It is as
though you had been introduced to heaven,
and then found yourself back on Earth. The
spirituality of a man cannot be completely
separated from the physical."

—Jacques Cousteau

Race to the Poles

In the era of Jules Verne, when few distinguished between pure science and sheer adventure, fanatical explorers vied to reach the earth's most remote locations. Their ostensible aim was scientific, but in many cases the quest for knowledge seems to have mushed in the tracks of the quest for publicity

NORTH POLE

1909

Peary Reaches Pole—Says Peary

A U.S. Navy commander, Robert Peary was a complex, troubled man whose ego was as bristly as his beard. Determined to be first to reach the Pole, he claimed he did so on April 6, 1909, in company with his associate Matthew Henson and four Eskimos. The problem: by his own account, he and his party would have had to cover 53 miles a day by dogsled for six days on their mad-dash round trip—a superhuman feat. David Roberts, author of *Great Exploration Hoaxes,* brands Peary's claim false.

1893-96
Nansen's Expedition Probes the Far North

Norwegian Fridtjof Nansen wedged his ship, the *Fram,* in pack ice and drifted from New Siberian Island off the Russian coast to Spitsbergen, Norway. In 1895 Nansen left the ship and attempted to reach the Pole by dog team, right. He failed but reached farther north than anyone before him.

1897
Andrée's Balloon Expedition Fails

How to access the inaccessible? Swedish explorer Salomon Andrée disappeared after setting out in a balloon, left, from Danes Island, Spitsbergen, hoping to drift across the Pole to America. The frozen bodies of the three expedition members were found 33 years later on White Island in the Northwest Territories. In 2000, Briton David Hempelman-Adams achieved Andrée's goal, even using a replica of Andrée's wicker basket.

1926
Byrd Flies over Pole—Says Byrd

In a rivalry that matched that of Peary and Cook, U.S. Navy officer Richard Byrd, above, vied with Norwegian explorer Roald Amundsen to be first to fly over the Pole. Byrd claimed he and associate Floyd Bennett succeeded on May 9, 1926. The two enjoyed a heroes' welcome in the U.S., but most historians believe their claim is false and that Amundsen's flyover by dirigible three days later was successful.

1908
Cook Reaches Pole—Says Cook

Physician Frederick Cook, above, was a former associate—and then a bitter rival—of Robert E. Peary's. When Peary returned from his April 1909 attempt to reach the Pole, he was dismayed to find that the charismatic Cook was claiming to have reached the goal a year earlier. Few experts believe Cook's claim today—he also claimed, falsely, to have been the first to summit Mount McKinley—but at the time the public supported him.

H. PONTING/ROYAL GEOGRAPHICAL SOCIETY

1911
Amundsen Is First to Reach the Pole

Echoing the Peary-Cook rivalry over the North Pole, Norwegian explorer Roald Amundsen, left, and Briton Robert Scott vied to be first to reach 90° south. Scott came close, but failed, during a 1901-04 expedition. In 1909 both men launched fresh expeditions for the goal. Amundsen was better supplied and his base was closer: he reached the Pole on Dec. 14, 1911. On Jan. 17, Scott and his team arrived at the Pole—only to find the Norwegian flag left by Amundsen, below.

milestones

ROYAL GEOGRAPHICAL SOCIETY

1907
Shackleton Fails

Briton Ernest Shackleton, above, was a colleague of Robert Scott when Scott attempted to reach the South Pole in 1901. The expedition failed, and Shackleton launched his own attempt six years later. His starving, frostbitten group turned back only 100 miles short of the prize.

1912
Scott's Team Perishes

Robert Falcon Scott, left, was short of everything on his mad dash to reach the Pole before his rival Amundsen—short of supplies, short of dogs (he used ponies, which were unsuited to the clime) and short of time (he finished a clear second). But he was long on courage, as he proved on the failed, nine-week return journey on which he and his four colleagues starved to death. His journal, later retrieved, is a classic in the annals of exploration.

H. PONTING/ROYAL GEOGRAPHICAL SOCIETY

1914-15
Endurance in the Ice

Ernest Shackleton's journey with 27 men on the aptly named *Endurance* has attracted a flurry of attention in recent years. When the ship was marooned in pack ice, his team wintered over. When it sank, Shackleton and five men sailed 800 miles in lifeboats through 20-ft. waves to reach South Georgia Island, then returned to the camp; all 28 men who began the expedition survived.

Sink Tank

The curator of ornithology at the New York Zoological Society was widely regarded as a bit odd, if undeniably brilliant. Back in the '30s, William Beebe was predicting that one day, "a human face will peer through a tiny window and signals will be passed back to the earth, such as 'We are above the level of Everest' and 'Clouds blot out the earth.'"

When he wasn't holding forth on the future of space flight (as he did in 1934), the specialist on birds was obsessing about how to get an up-close look at inner space: the bottoms of the oceans. He was fascinated by references to men's exploring the sea floor in *The Iliad* and in the ancient story of the Athenian siege of Syracuse. He had even carried on a running debate years earlier with fellow Zoological Society member Theodore Roosevelt about how to get there: Beebe envisioned a craft in the form of a cylinder; T.R. favored a metal sphere. By the early 1930s, Beebe had been won over to the spherical design—inventor Otis Barton had demonstrated that a heavy steel globe could best withstand the crushing pressure exerted by seawater at depths of more than a few hundred feet by distributing stress equally throughout the structure.

The two fine-tuned Barton's craft on Nonsuch Island (just off Bermuda) and conducted several test dives, each probing deeper than the last. Finally, on Aug. 15, 1934, they descended to 3,028 ft. The descriptions they sent back through a telephone cable put surface dwellers on notice that an entirely new world (containing hundreds of previously unknown species of fish and plant life) had been broached at last: "Strange, ghostly, dark forms hovering in the distance," they reported. "They look like stars gone mad." Beebe and Barton may have only dipped mankind's toe into the upper layer of the oceans, but their work opened up a vast new arena for scientific exploration.

Photograph by William Beebe—National Geographic Society

MEN FROM INNER SPACE Beebe and Barton in 1930, in the early days of their work. Their 5,000-lb. submersible was equipped with a 3-in.-thick quartz glass window and a 3,500-ft. cable

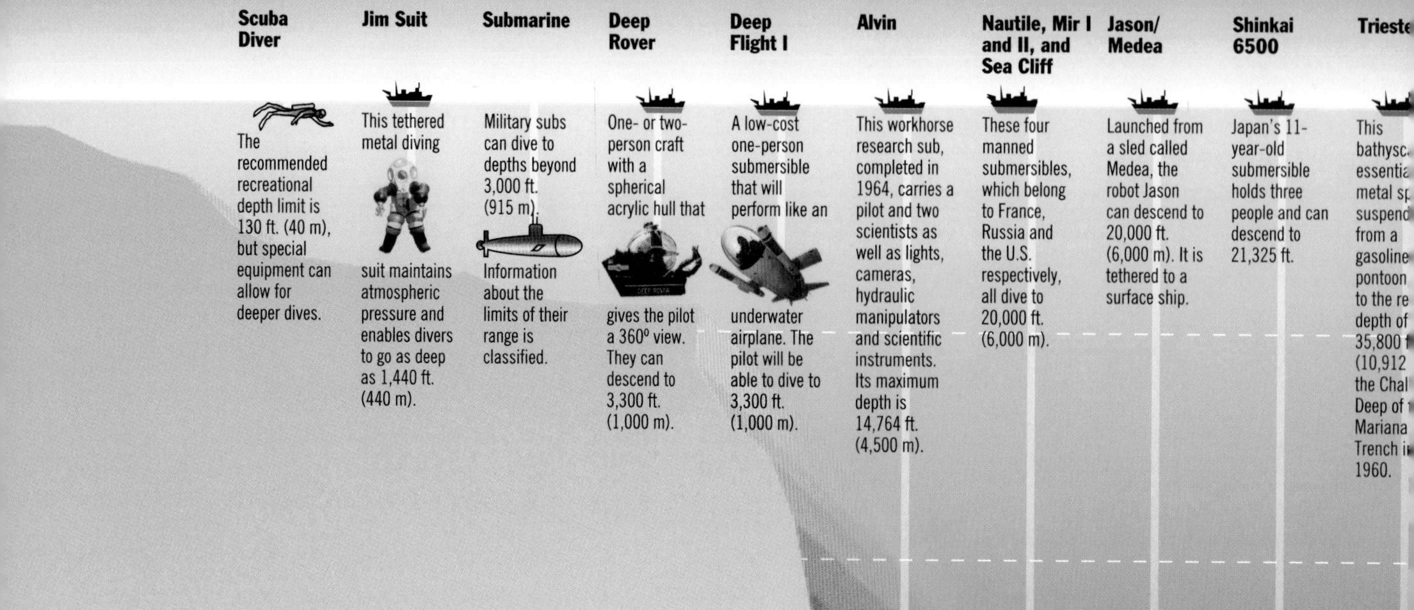

Scuba Diver	Jim Suit	Submarine	Deep Rover	Deep Flight I	Alvin	Nautile, Mir I and II, and Sea Cliff	Jason/Medea	Shinkai 6500	Trieste
The recommended recreational depth limit is 130 ft. (40 m), but special equipment can allow for deeper dives.	This tethered metal diving suit maintains atmospheric pressure and enables divers to go as deep as 1,440 ft. (440 m).	Military subs can dive to depths beyond 3,000 ft. (915 m). Information about the limits of their range is classified.	One- or two-person craft with a spherical acrylic hull that gives the pilot a 360° view. They can descend to 3,300 ft. (1,000 m).	A low-cost one-person submersible that will perform like an underwater airplane. The pilot will be able to dive to 3,300 ft. (1,000 m).	This workhorse research sub, completed in 1964, carries a pilot and two scientists as well as lights, cameras, hydraulic manipulators and scientific instruments. Its maximum depth is 14,764 ft. (4,500 m).	These four manned submersibles, which belong to France, Russia and the U.S. respectively, all dive to 20,000 ft. (6,000 m).	Launched from a sled called Medea, the robot Jason can descend to 20,000 ft. (6,000 m). It is tethered to a surface ship.	Japan's 11-year-old submersible holds three people and can descend to 21,325 ft.	This bathyscaphe, essentially a metal sphere suspended from a gasoline pontoon, to the record depth of 35,800 ft (10,912 m) the Challenger Deep of the Mariana Trench in 1960.

■ deep sea

Probing the Abyss

The world beneath the ocean waves remains a great frontier whose rewards could be enormous: oil and mineral wealth to rival Alaska's North Slope and California's Gold Rush streams; scientific discoveries that could change our view of how the planet—and the life-forms on it—evolved; natural substances that could yield new medicines and new classes of industrial chemicals.

Getting there, though, forces explorers to cope with an environment just as perilous as outer space. Unaided, humans can't dive much more than 10 ft. before increasing pressure starts causing pain in the inner ear, sinuses and lungs. Frigid subsurface water rapidly sucks away body heat. And even the most capacious of lungs can't hold a breath for more than two or three minutes.

For these reasons the modern age of deep-sea exploration had to wait for two key technological developments: engineer Otis Barton's bathysphere—essentially a deep-diving tethered steel ball—and the invention of scuba in the 1940s by Jacques Cousteau and Emile Gagnan. Barton's bathysphere wasn't maneuverable: it could only go straight down and straight back up again. Swiss engineer Auguste Piccard solved the mobility problem with the first true submersible, a dirigible-like vessel called a bathyscaphe, which consisted of a spherical watertight cabin suspended below a buoyant gasoline-filled pontoon.

In 1960 Piccard's *Trieste* took U.S. Navy Lieut. Don Walsh and Piccard's son Jacques 35,800 ft. down beneath the Pacific to the Challenger Deep in the Mariana Trench. After its success, the number of submersibles expanded dramatically. The Woods Hole Oceanographic Institute's longtime workhorse, the three-person *Alvin*, was launched in 1964. The first tethered robots—the so-called remotely operated vehicles, or ROVs—were developed several years later. The result was a remarkable period of underwater discovery that transformed biology, geology and oceanography. Today scientists view the deep sea as an area constantly reformed by tectonic and volcanic activity and filled with exotic life-forms whose properties have yet to be explored.

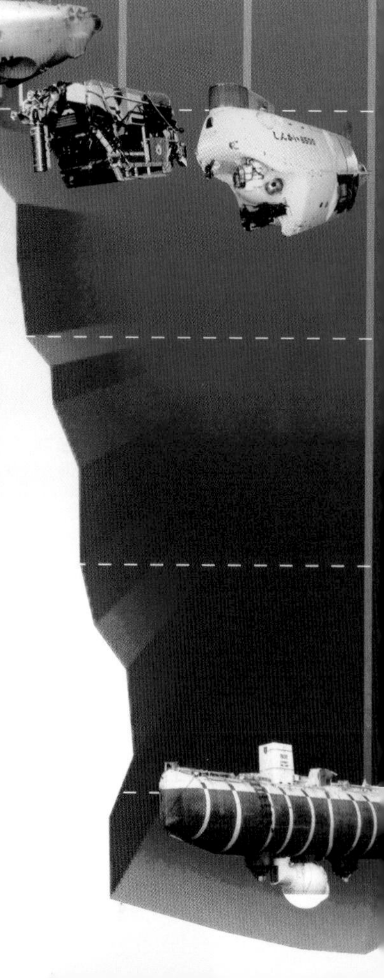

WILD BLUE UNDER The illustration at right offers a visual chronicle of the milestones in technology that have provided us with new views of a vast, surprising world

Sea Cliff

Amphipod	Clams	Hydrothermal Vents	Manganese Nodules	Sperm Whale	Tubeworms	Deep-Sea Cod	Giant Squid	Plankton

... stage
...hat is
...ed to the
...e, Kaiko
...ed the
...nger
... in 1995.
...ped at
...B ft.
...1 m), an
... length
...of the
...e's
...l.

Amphipod
Proof that there is life at the very bottom of the sea, this 2-in. (5-cm) crustacean was found 36,200 ft. (11,030 m) down in the Mariana Trench.

Clams
Colonies of these mollusks grow near life-sustaining hydrothermal vents. The deepest known population of clams was found off Japan at a depth of 20,886 ft. (6,366 m).

Hydrothermal Vents
These superheated structures form in regions where the sea floor is spreading, at depths anywhere from 2,600 ft. to 20,000 ft. (800 m to 6,000 m).

Manganese Nodules
Commonly found at depths of 14,000 ft. to 17,000 ft. (4,270 m to 5,180 m), these potato-size nuggets are rich in iron, copper, cobalt, manganese and nickel.

Sperm Whale
The world's deepest diving mammal, it weighs as much as 60 tons and has been recorded at depths as great as 8,000 ft. (2,440 m).

Tubeworms
Found only near vents in the Pacific Ocean at an average depth of 7,300 ft. (2,225 m), these 8-in.-(0.2-m-) long organisms have no mouths.

They get their energy from bacteria that live in their stalks.

Deep-Sea Cod
These fish, found at 4,600 ft. (1,400 m), resemble their shallow-water relatives, but

have larger eyes and an enlarged dorsal fin that acts like a sensory antenna.

Giant Squid
Believed to grow up to 64 ft. (20 m) long and live as deep as

3,300 ft. (1,000 m), these "monsters of the deep" have never been seen in their natural habitat.

Plankton
Eaten by anything from barnacles to humpback whales, these microscopic animals and plants dwell in the well-lighted top 660 ft. (200 m) of the ocean. However, they can sometimes be found as deep as 1,000 ft. (300 m).

5,000 ft.

10,000 ft.

15,000 ft.

20,000 ft.

25,000 ft.

30,000 ft.

35,000 ft.

NORBERT WU

UNDERWATER, A NEW WAVE

The weird-looking submersible at right is *Deep Flight 1*, a 14-ft. long, 2,900-lb. vehicle that took its maiden voyage in the mid-1990s. Its pilot lies prone in a body harness, his or her head protruding into the craft's hemispherical glass nose.

Deep Flight is only one of a new armada of undersea explorers that debuted in the '90s. Among others was Japan's *Kaiko*, a 10.5-ton remotely operated vehicle that touched down in the Challenger Deep of the Mariana Trench in 1995, where it found a sea slug, a worm and a shrimp, proof that even one of the most inhospitable places on earth is home to complex life-forms.

On its very first series of missions in 1991, Japan's *Shinkai 6500* submersible found unsuspected deep fissures on the edge of the Pacific plate, one of three tectonic plates that converge under the southern part of the island nation.

The cheapest way to explore the oceans may be the new free-floating autonomous underwater vehicles, or AUVs. These untethered craft can roam the depths without human intervention for months on end, accumulating data.

Columbus of the Silent World

Inventing the Aqua-Lung to defy the barriers that kept us from probing our own planet, Jacques Cousteau gave mankind a brave new world to explore

THE WATER MAY BE THE TRANSLUCENT surf off Bermuda, an ice-skimmed quarry in Vermont, the Pacific rolling in majestic rhythm toward the shores of San Diego. Around the world, divers spend hours deep beneath the waves, sprites soaring in an alien realm. That humans have been able to discover a world long inaccessible to man is largely because of the work of a single visionary Frenchman, Jacques-Yves Cousteau—at once the pioneer, prophet, poet and promoter of the deep. As the developer of the Aqua-Lung, he set divers free to roam in the kingdom of the fish. As captain of the research vessel *Calypso*, he became a noted international exponent of conservation. As an author, filmmaker and expert in underwater photography, he was the dean of undersea explorers.

Young Cousteau graduated second in his class from France's naval academy, and after a car crash left his arms badly injured, he spent hours working their strength back by swimming daily in the Mediterranean. In 1935 a fellow naval officer gave him a pair of goggles used by pearl fishermen. Cousteau put his head beneath the surface of the sea and, as he later told it, his life seemed to change almost instantly: "There was wildlife, untouched, a jungle at the border of the sea, never seen by those who floated on the opaque roof."

Cousteau explored this jungle, but he was still tethered to the surface by the need for air. Looking for a way to supply it, he tried an oxygen lung based on a design developed by the British as early as 1878. He almost killed himself. He did not know the fatal flaw of oxygen: it becomes toxic at depths below 30 ft. Twice he had convulsive spasms, and was barely able to drop his weights and return to the surface.

Cousteau allowed World War II to distract him only briefly from his search. Under the eyes of the indifferent Germans in Occupied France, he worked with a brilliant engineer named Emile Gagnan to develop a lung that would automatically feed him safe, compressed air at the same pressure as the surrounding water, freeing him to swim with both arms. One day in 1943 he waddled out into the Mediterranean under the 50-lb. Aqua-Lung and realized his dream. He was free. "Delivered from gravity and buoyancy, I flew around in space," he said.

After the war, Cousteau sold the French navy on the virtues of the Aqua-Lung and got leave to conduct oceanography sponsored by the government aboard the 360-ton *Calypso*, a refitted World War II British Royal Navy minesweeper. Aboard her, he gathered the material for the books and films that would bring him fame, like 1953's *The Silent World*, his first book, in English, which sold more than 5 million copies.

Over the years, his rewards were many: expeditions to study waterways all around the globe, more than 100 undersea documentary films, more than two dozen books, three Oscar awards, induction into the Académie Française. But the great discoverer also knew tragedy. His son and heir apparent Philippe died in a plane crash in 1979. Later Cousteau parted ways with Jean Michel, his eldest son and close collaborator; their dispute ended up in court.

As his knowledge of the oceans grew, Cousteau underwent a final transformation: he became the tribune of the undersea world. Before his death in 1997, the man who began his career as the Columbus of the deep spent his autumnal years in a different role: as an aging King Canute, defiantly raging against the rising tide of human exploitation of the seas.

THE ESSENTIAL COUSTEAU

THE SILENT WORLD, 1953
Cousteau's first best-seller is a fine primer on the undersea environment

THE LIVING SEA, 1963
In this book Cousteau brings us up-to-date on his research exploits

THE SILENT WORLD, 1956
His first film won awards and pioneered underwater cinema

WORLD WITHOUT SUN, 1964
Innovations in under-water filming made this superior to his earlier movies

THE UNDERSEA WORLD OF JACQUES COUSTEAU, 1968-76
His long-running TV series brought Cousteau his biggest audience

THE OCEAN WORLD, 1985
This oversize book is an eye-popping introduction to the marvels of the undersea kingdom

THE COUSTEAU SOCIETY
Founded by Cousteau in 1973 to promote his message, this group claims more than 150,000 members worldwide. It operates a website at www.cousteausociety.org

Full Fathom Five

It was a surreal image: "metal biscuits with ears." That's how a young Turkish sponge diver from a small Mediterranean village described some curious objects he had spotted lying near a sunken shipwreck. When George Bass, a nautical archaeologist who had been rummaging around the floors of the Mediterranean coast for 25 years, heard that description in the summer of 1982, he thought—he hoped—that it might put him on on to something.

That something turned out to be the earliest intact shipwreck recovered to that time, a fully laden cargo vessel that had gone to its watery grave perhaps 3,400 years ago, about the time King Tut occupied the throne in Egypt. The discovery was located near the town of Kas, less than 100 yds. off the jagged, arid southern Turkish coastline and more than 145 ft. below the surface.

The wreck yielded a rich trove of Bronze Age artifacts, some of which are now at a museum in Bodrum, Turkey: 6,000 lbs. of copper ingots (the "biscuits"), glass ingots, a store of tin (which was combined with copper to make the bronze that gives the era its name), pottery, gold objects, amphoras filled with glass beads, some ivory from an elephant tusk and a hippopotamus tooth.

The ship is about 65 ft. long, rigged for a single square sail. Apparently it foundered on the coast's treacherous rocks and went straight down, thus saving much of its cargo. The discovery of a small seal, no larger than a button, with markings similar to those used by the Greek merchants who dominated the ancient Mediterranean trade routes, helped provide a possible nationality for the vessel.

Reflecting on the science of nautical archaeology he helped pioneer, Bass said, "[Before the find] we knew more about the safety pins and sewers of Athens than we did about the ships that made Athens great."

Photograph by William Curtsinger—National Geographic Society

WRECKED Because of the great depth of the find—145 to 170 ft.—divers could make only two brief trips per day. The pressure was so disorienting, George Bass said, "it was like working down there on three martinis"

AN UNDERSEA INDIANA JONES

In 1960, about 12 years after Jacques Cousteau and Emile Gagnan pioneered scuba diving, George Bass began combining the new undersea technology with his scientific discipline, archaeology. Result: nautical archaeology, which has proved a potent new tool to uncover artifacts from the depths of the ocean—and the past.

Bass adapted long-standing archaeological surveying techniques to underwater work, like roping off the search area into a grid for systematic searching. He also pioneered the use of an underwater "telephone booth" to help divers talk to the surface.

Below are several amphoras discovered at the site off Turkey; a cache of cobalt-blue glass ingots is the earliest such glass ever found.

WILLIAM CURTSINGER/NATIONAL GEOGRAPHIC SOCIETY

RALPH WHITE—CORBIS

WOODS HOLE OCEANOGRAPHIC INSTITUTE

WOODS HOLE OCEANOGRAPHIC INSTITUTE

FRAGMENTS OF A DISASTER

CHINA
Debris was found across a huge field, but these china plates went to the bottom in orderly rows

DECK CHAIR
Its metal framework survived, but marine creatures had eaten most of the ship's woodwork

PURSER'S SAFE
Despite the hopes of treasure hunters, no valuables were recovered from the wreckage

JONATHAN BLAIR—NATIONAL GEOGRAPHIC SOCIETY

"BOTTOM GUN"
Ballard aboard the submersible *Delta*.
The U.S. Navy helped fund the search
for the *Titanic*; afterward, Navy
skipper John Lehman gave him a
hat reading: "Bottom Gun"

■ **deep sea**

Tragedy's Trove

For 22 days in August 1985, the French vessel *Le Surôit* had "mowed the lawn" in the North Atlantic Ocean, roaming over a 150-sq.-mi. target area with a sonar device that provided high-resolution maps of the seabed. But after surveying 80% of the expanse, *Le Surôit* had found no trace of its prey. Then the U.S. Navy research vessel *Knorr* began combing the remainder of the target area with a sonar-and-video platform called Argo, which was towed behind the ship at a depth of 12,500 ft.

Early in the morning of Sept. 1, the Argo detected something on the seabed. Was it the remains of a huge steam boiler? "That's it!" cried expedition leader Robert Ballard, a veteran undersea explorer. He and his team had found the legendary White Star liner *Titanic*, which broke apart and sank in the early-morning hours of April 15, 1912, after hitting an iceberg, taking some 1,500 people with her to the bottom.

Eleven months later, Ballard and his team returned and explored the remains of the great ship using two devices, the submersible *Alvin* and the remotely operated vehicle Jason Jr. The two craft toured the wreckage of the luxury liner, wandering across the decks past corroded bollards, peering into the officers' quarters and through rust-curtained portholes where doomed passengers stood 74 years before. Ballard found no trace of the rumored 300-ft. tear supposedly left in the ship by the iceberg, leading him to believe only a small rip sank the ship. And he theorized that *Titanic* broke apart not when she hit bottom, but as she sank: the stern, which settled on the bottom almost 1,800 ft. from the bow, had swiveled 180° on its way down. Both theories were reflected in the 1999 hit movie *Titanic*, which employed recent underwater photography of the ship.

Photograph by Emory Kristof—National Geographic Society

"RUSTCICLES" Following Ballard's success, other expeditions have explored *Titanic's* remains. The main picture was taken by the submersible *Mir 1* in 1991 and shows the bow railing of *Titanic* behind the forward anchor crane. The slant of the wreck's "rustcicles" shows the direction of the current

KON-TIKI MUSEUM—LIAISON

THOR HEYERDAHL 1914-

1930s
While living in the South Pacific to complete his doctorate in zoology, Heyerdahl becomes convinced that ancient island inhabitants must have come from South America

1947
Heyerdahl and a crew of five shock the world by successfully rafting from South America to Polynesia

1960s
Excavates the giant statues on Easter Island

1969-70
Sails twice from Morocco to Barbados aboard *Ra,* a papyrus boat based on ancient Egyptian drawings

1977
Builds a reed boat to research ancient trade routes in the Middle East

1990
Leads a campaign to save ancient step pyramids in the Canary Islands similar to those of Egypt and Mexico

1992
Thousand-year-old paintings depicting large oceangoing vessels are discovered in a Peruvian temple, further supporting Heyerdahl's theories

Floating a Theory

In the 1930s, most scientists who studied the movement of early man around the globe believed that the Polynesian islands of the South Pacific had first been settled by seafaring Stone Age peoples from South East Asia. Yes, there were a few problems with that model—like the fact that ancient Asians would have had to fight prevailing east-to-west currents and winds for 10,000 miles to get there, or that the first Polynesians grew plants native to South America, or that their language contained words suspiciously similar to an ancient Peruvian language. But that still didn't give an upstart young scientist who wasn't even trained in archaeology the right to thumb his nose at the prevailing orthodoxy. That didn't stop Norwegian scholar Thor Heyerdahl. Working on his Ph.D. in zoology on the remote Pacific island of Fatu-Hiva, young Heyerdahl came to believe that the first South Seas inhabitants must have traversed the ocean from South America.

In 1947 Heyerdahl decided to test his theory by taking it out of the academy and into the deep blue sea. Lashing together a raft from balsa logs and employing techniques known to have been used by the ancient Peruvians, Heyerdahl and five shipmates (none of whom had any previous experience in sailing) built a 35-ft. raft they christened *Kon-Tiki* and set out for Polynesia. After 101 days at sea, they landed on an uninhabited island in the Raroia atoll group. Heyerdahl's book about his voyage became a best seller and a high school classic. And the world (if not the entire scientific community, which does not lightly forgive being made to look foolish) was at last convinced that South American peoples may indeed have crossed the Pacific thousands of years ago.

Photograph: Kon-Tiki Museum—Liaison

DEEP BLUE ZOO The *Kon-Tiki* (a fusion of the ancient Peruvian word for sun and the name of a chief said to be descended from the sun god) was shadowed by dolphins, sharks, whales and several never-identified creatures, such as a giant glow-in-the-dark fish the sailors estimated to be 30 ft. long

To the Ends of the Earth

When soaring into the stratosphere higher than any man had ever voyaged before became old hat, Auguste Piccard took to the sea—and dived lower

Men began sailing into the skies below hot-air balloons in the age of the Montgolfier brothers in the 1700s, but the first such aviators very quickly came up against one of nature's brick walls: somewhere between 14,000 and 15,000 ft., the air starts to become too thin to breathe. That wasn't a problem for sightseers or soldiers hoping to spy on enemy lines—for them, the best views are at far lower altitudes. But for a scientist trying to measure the trace amounts of cosmic radiation that continually bombard the earth (but almost all of which is dissipated before reaching the ground by the earth's thick, lower atmosphere), 15,000 ft. is far too low to achieve meaningful readings.

Enter Swiss-born physicist Auguste Piccard, who had been captivated since boyhood with the potential of a pressurized metal sphere to take human explorers to places they could otherwise never go. As a child, Piccard dreamed of using such a craft to explore the bottoms of the oceans, but after receiving his degree in physics, he began collaborating with Albert Einstein to design detection equipment sensitive enough to measure cosmic radiation. Still, the devices they created were useless unless they could somehow be taken into the upper atmosphere.

So with the help of Belgian brewery engineers (who were skilled in working with aluminum, a lightweight and electrically neutral material that could carry instruments aloft with a minimum of interference), Piccard fabricated an airtight gondola that would maintain an interior pressure similar to that at sea level, and suspended it beneath a hydrogen balloon. After a number of trial flights (each of which ascended to a new record height), Piccard was ready to slip the surly bonds of earth. On May 27, 1931,

Piccard and his assistant, Paul Kipfer, took off from Augsburg, Germany, and reached the stratosphere in less than 30 minutes. They topped out at slightly more than 56,000 ft., where the air pressure is less than one-tenth its density on the earth's surface. The good news was that Piccard and Kipfer experienced no ill effects from the altitude. The bad news was that the gas valve they used to control their altitude broke, leaving them stranded in the stratosphere. After wandering for 17 hours in the upper atmosphere, they were finally able to bring the balloon to a controlled landing on a glacier in the Alps. Later flights would ultimately carry Piccard to a height of 61,237 ft.

Having gone higher than any human before him, Piccard set his sights in the late 1930s on his original goal: going lower than anyone had ever dared imagine. He began developing a watertight diving craft to explore the oceans at depths that would have crushed submersibles built up to that time. His first bathyscaphe—essentially a reverse gondola, designed to keep pressure out, rather than in—was completed (after a forced hiatus for World War II) in 1948 and tested at great depths off the coast of Senegal, under the command of a team led by Jacques Cousteau. In 1953 young Jacques Piccard joined his father, and they descended together to a record depth of 10,300 ft. in the bathyscaphe they christened *Trieste*.

After Auguste Piccard retired, Jacques and U.S. Navy lieutenant Don Walsh dropped to 35,800 ft. (more than seven miles) in the Pacific Ocean's Mariana Trench. The family's third generation returned to the skies: in March 1999, Bertrand Piccard, Jacques's son, spent 20 days aloft to complete the first nonstop, round-the-world balloon flight.

UP, UP AND AWAY!

1783
The Montgolfier brothers make the first hot-air balloon flight. Although tethered to the ground by a long rope, the brothers hover over Annonay, France, at 6,600 ft.

1785
Jean-Pierre Blanchard of France and John Jeffries, an American, become the first to cross the English Channel in a hot-air balloon

1859
John Wise flies from New York City to St. Louis in the first air-mail balloon flight

1861
In the first use of balloons in war, Thaddeus Lowe telegraphs Confederate positions to Union troops

1897
Swedish scientist Salomon August Andrée is lost while seeking the North Pole in a balloon

1931
Auguste Piccard and Paul Kipfer are the first human beings to reach the stratosphere

LATE 1940s
The Air Force's top-secret Project Mogul uses high-altitude balloons to spy on Soviet atomic tests

1950s
James Van Allen studies cosmic rays with "rockoons"

1961
Malcolm Ross and Victor Prather ascend to 113,740 ft., still the altitude record (excluding space flight)

大石下り2

5265

大石

MYSTERY OF THE FAULT LINE

The San Andreas Fault is the fissure where two enormous pieces of the earth's crust, the North American Plate and the Pacific Plate, are constantly sliding past and grinding against each other. (Above, a view of the fault in the Carrizo Plain, 100 miles northwest of Los Angeles.) The fissure breeds earthquakes, from the 1906 giant that ravaged San Francisco to the 1994 killer that struck Northridge, a suburb outside Los Angeles.

The fault has bred a mystery as well: geologists are baffled as to why it is not generating the high heat underground that they expect to see in a region of so much stress. Extreme heat is present in most other faults; a law Byerlee's Law, named for James Byerlee of the U.S. Geological Survey) is used to predict it. Yet six miles underground, the heat beneath the San Andreas Fault is "only" 300°C; Byerlee's Law says it should be 400°. The cause of the depressed temperature is one of the great unresolved questions of contemporary geology.

Getting the Drift

On Jan. 17, 1995, a massive earthquake rocked Kobe, Japan, as a 7.2 tremor—lasting only 20 seconds—ravaged the island nation's sixth largest city, collapsing houses, wrecking roadways, igniting fires, destroying ports and leaving more than 6,000 dead. Insurance companies describe such quakes as "acts of God"; scientists regard them as a by-product of deep geological processes in which giant plates—the continents—slowly move across the planet's mantle, the deep area between the earth's rigid crust and its molten core.

Continental drift: when German meteorologist Alfred Wegener first proposed this mind-boggling idea in 1912, many geologists regarded him as if he were a member of the Flat Earth Society. Convinced that the continents were anchored firmly in place, they dismissed his theory that the earth's major land masses had once been huddled together in a single supercontinent, which he called Pangaea (Greek for "whole earth"), then began slowly drifting apart—and were still on the move. Wegener had plenty of evidence, ranging from the jigsaw-like fit of the continents to the discovery of matching fossils on opposite sides of oceans, but he was unable to provide an explanation of the mechanics that drove the global breakup.

For years Wegener and continental drift were held up to derision—until scientists in the 1960s found plausible evidence of the process at the bottom of the seas, where crustal plates were pulling apart and lava was welling up to form a new sea floor. Suddenly, Wegener's disreputable ideas became reputable. Renamed plate tectonics, they gave geology a single unifying theory, linking earthquakes and volcanoes to the formation of mountain ranges and ocean basins. Sadly, Wegener, who perished on the Greenland icecap in 1930 at age 50, didn't live to see his thesis accepted as the most profound discovery in modern geology.

Photograph: Robert Patrick—Corbis Sygma

TWISTED An elevated train track in Kobe was manhandled by the 1995 earthquake. Such events are all too common on the Japanese archipelago, which sits astride a geological fault line in the Pacific Ocean

Thar She Blows!

There is but one enigma that every volcanologist since Pliny the Elder (who died while observing the eruption of Mount Vesuvius in A.D. 79) has wanted to resolve: how to know when a volcano is going to blow.

Since Pliny's time, scientists have made it about halfway to their goal. Volcanologists now make some calls with stunning accuracy, as they did in 1991, when they convinced Philippine government officials to evacuate 35,000 people from the area around Mount Pinatubo two days before it erupted, right. But scientists still sound the occasional false alarm: in the mid-'70s, at the urging of volcano experts, officials on Montserrat evacuated 70,000 people from the base of the Soufrière volcano and kept them away for more than three months, but the volcano didn't erupt for another 20 years. And they miss some big ones entirely: witness Indonesia's Mount Merapi, which erupted without warning in 1994.

Most tragic of all, however, are the cases in which scientists sound the alarm and are ignored, only to be proved right. In 1985 Colombian seismologists warned their government that the Mount Ruiz volcano was smoldering dangerously. Their data were too spotty to convince officials, however; nothing was done. One month later, the mountain erupted, claiming 23,000 lives.

Modern science has advanced in giant strides beyond that of the ancients, yet our continuing inability to predict eruptions can be deadly. In 1993 a team of 12 scientists ventured to the crater's rim at the top of Colombia's ominously rumbling Galeras volcano, hoping to get a more accurate fix on when the next eruption would take place. Without warning, the mountain exploded, sending comets of lava and rock the size of cars soaring into the sky at faster than the speed of sound. Half of the team was killed instantly, and the six survivors (all seriously wounded) learned firsthand how far we have yet to go in fulfilling Pliny's quest.

Photograph by Alberto Garcia—Corbis/Saba

ON THE RUN Journalists waiting to record the eruption of Mount Pinatubo in the Philippines in July 1991 tarried a bit too long—and raced to flee its cloud of ash

PREDICTING ERUPTIONS

How do scientists try to forecast a blow? Instruments placed on the sides of a volcano, within its cone or mounted on airplanes can detect tiny changes in tilt, the composition of gas emissions and gravity (as magma surges upward under high pressure, the mountain becomes more massive and thus exerts an infinitesimally stronger gravitational tug). Measurements taken by satellite can also gauge harmonic tremors—rhythmic shock waves that reverberate through a volcano prior to a blow. Ground-based lasers can pick up minute deformations in the mountain's shape that are invisible to the naked eye.

Experts predicted some aspects of the 1980 eruption of Mount St. Helens in Washington State, shown below, but were caught off guard by both its fury and the extent of the mudflows it generated.

DAVID WEINTRAUB—PHOTO RESEARCHERS

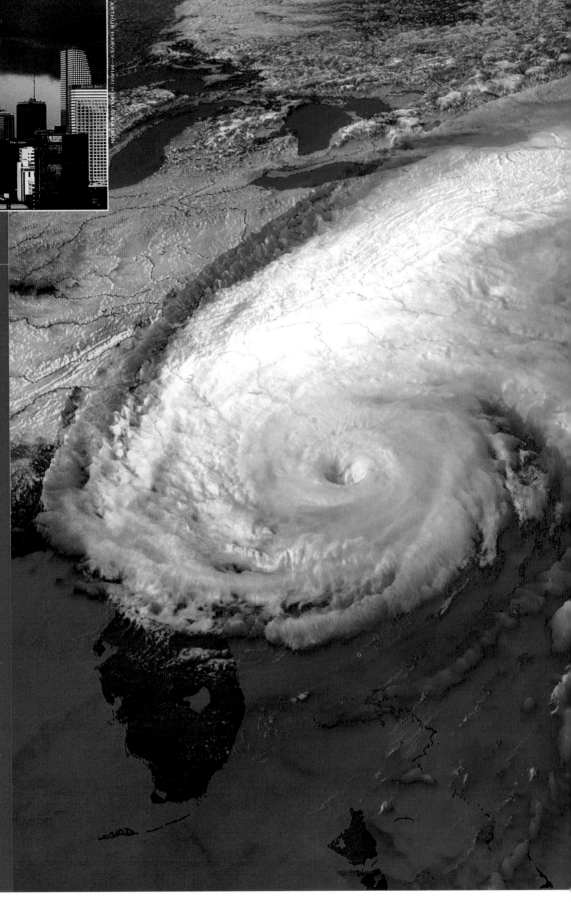

A NEW TWIST
IN FLORIDA

We think of "Tornado Alley" in the Midwest as the home of big twisters, but the funnel-shaped storms can be found just about anywhere—like the one above that stunned Miamians when it touched down in the middle of the city on May 12, 1997.

Scientists at the National Severe Storms Laboratory in Norman, Okla., have developed new tools to understand and predict tornadoes. To explore their nature, they created "turtles," canisters weighted with lead and packed with temperature and pressure gauges, then placed in the path of an oncoming tornado. To predict the twisters' movement, they began using Doppler radar, which makes use of the fact that radio waves shift frequency depending on whether the objects they bounce off—in this case, water drops inside storm clouds— are advancing or receding. As the winds inside the clouds begin to spin, the droplets show up on radar screens as tighter and tighter swirls. Speeding up storm warnings, Doppler radar is a lifesaver.

Warm-Up Act

The picture at left shows 1999's Hurricane Floyd, a weather system so vast it blanketed the East Coast of the U.S. from Florida to Maine. The image is based on data from the National Oceanographic and Atmospheric Administration satellite GOES-8. The U.S. deploys two types of satellites to monitor storm systems: Geostationary Operational Environmental Satellites (GOES), which maintain a fixed position 22,300 miles above the equator, and Polar-orbiting Operational Environmental Satellites (POES), which make 14.1 north-to-south orbits of the earth each day. But when a hurricane is raging, the best way to measure its power at a given moment is to fly right into it. The NOAA's "hurricane hunter" aircraft do just that, deploying capsules—dropwindsondes (for drop wind sounding)—that measure barometric pressure, temperature, water vapor and wind speed as they fall through the atmosphere, sending data back to the aircraft and on to the National Hurricane Center, which uses the information to create forecast models.

Hurricane Floyd did America a favor: it veered away from the coast, though its potent punch still left some 70 dead. The bad news is that scientists believe Hurricane Floyd is nothing compared with what may lie ahead. In the century that's just begun, says M.I.T. atmospheric scientist Kerry Emanuel, global warming may breed "supercanes" that far exceed a contemporary hurricane's upper limit of 180 m.p.h.—more than capable of sending a 30-ft. wall of water surging inland, flattening houses and inundating coastal cities. Atmospheric pollution and the greenhouse effect are expected to heat not just the air but also the surface of the oceans, and it is the thermal energy of that water that fuels typhoons and hurricanes. Emanuel estimates that sustained winds in future hurricanes could conceivably top 200 m.p.h.

Photograph: Hal Pierce/Fritz Hasler—NASA/Goddard Space Flight Center from NOAA GOES-8 Data

MUST TO AVOID Though predictions of Hurricane Floyd's potential for destruction had worried motorists clogging highways on the East Coast in September 1999, the storm proved less deadly than expected

Just Hanging On

The tourists who stood on the poop deck of the Russian icebreaker *Yamal* were looking for the North Pole—but they found the North Pond. On an August 2000 cruise, they were scheduled to crunch through 6 to 9 ft. of arctic ice and stand at the top of the world. But when the *Yamal* arrived at 90° north, there was no ice to crunch. In fact, there was a mile-wide lake with gulls circling overhead. Although it later turned out that the watery Pole was less unusual than some scientists had initially thought, it was a spooky addition to the accumulating mountain of evidence that the world is getting balmier.

The theory of global warming, as popularly understood, has two parts: the observation that average temperatures around the globe are rising steadily (if slightly) each year and the conclusion that the heat-up is caused by the pollutants that humanity has been pumping into the atmosphere since the dawn of the Industrial Revolution.

The problem with the theory involves cause and effect: although the earth is definitely getting warmer, it's hard to prove that this isn't due as much to natural, centuries-long cycles of warming and cooling as it is to pollution's kicking the greenhouse effect into overdrive. (The greenhouse effect is the process by which the earth's atmosphere traps, moderates and evenly distributes the heat of the sun.) And here's the problem with the problem: if we wait to find out for sure what's causing global warming, it will probably be too late to do anything about it.

In the meantime, the picture of the not-so-distant future isn't pretty: should present trends continue (regardless of their cause) and global sea levels rise just 3 ft. over the next century, we can look forward to worldwide floods, a dramatic increase in killer storms, the destruction of one-third of the world's forests, the devastation of important food sources like ocean fisheries and 100 million people made refugees.

Photograph by Flip Nicklin—Minden Pictures

OOMPH! A polar bear clambers onto the ice in Wager Bay, Canada. In 2000 sea ice in the Arctic was 40% thinner and covered 6% less area than in 1980. Over the same span, average temperatures climbed as much as 7% in Alaska, Siberia and parts of Canada

VAN BUCHER · PHOTO RESEARCHERS

Bermuda Triangle

How old are the stories of something spooky happening in the triangle formed by Bermuda, Puerto Rico and Florida? Well, Christopher Columbus said he witnessed "a great flame of fire" crash into the ocean there. In the following centuries, tales of large vessels vanishing circulated. The myth got its biggest boost in 1945, when the five planes of U.S. Navy Flight 19, above, disappeared. But the last word on what is (or isn't) happening in the Bermuda Triangle is perhaps best left to people who count money, rather than boats or airplanes. Norman Hooke, a spokesman for Lloyd's of London, says simply, "There are just as many losses in any other wide expanse of ocean."

What on Earth Were They Thinking?

Running around in circles—including crop circles—generations of explorers have searched for terra incognita, but the places they sought often have more to do with romance than science

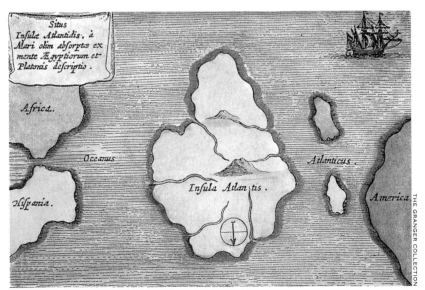

THE GRANGER COLLECTION

Atlantis

Plato mentioned Atlantis in two dialogues (the *Timaeus* and the *Critias*) and describes in some detail where it was—beyond the Pillars of Hercules, before "the opposite continent"—perhaps the Rock of Gibraltar and the Americas (left, a 1678 German map of the Atlantic). The gods destroyed the rich island nation with an earthquake and floods, he said, in either 8,570 B.C. (*Timaeus*) or 9,421 B.C. (*Critias*). So why haven't we found Atlantis? The two best theories are that we have and just don't know it (the Minoan cities of Akrotiri on Santorini and Knossos on Crete are good candidates)—and that Plato simply made the whole thing up.

Seven Cities of Gold

In 1539 Catholic cleric Marcos de Niza returned to Mexico City from the unexplored wilds to the north, claiming to have seen a kingdom, Cibola, where 10-story buildings were made of gold. The viceroy quickly appointed a young nobleman, Francisco Vásquez de Coronado, to lead the conquest of the kingdom that Niza, who perhaps had an overactive imagination, called the "Seven Cities of Gold." After a three-year trek up the Rio Grande Valley into present-day Arizona and New Mexico, Coronado came upon the village of Hawikuh, inhabited by the Pueblo indians called Zunis. Their huts were made of mud—by coincidence, Friar Niza's new name.

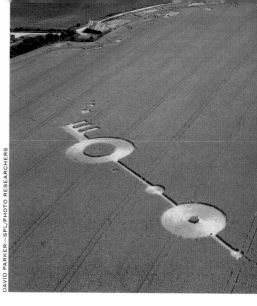

Crop Circles

When mysterious patterns showed up in the fields of Wiltshire, England, in the mid-'80s, nobody other than annoyed local farmers took much notice. But then in 1989, a book on the phenomenon, *Circular Evidence,* became a best seller. Among the theories that tried to explain them: variations in the earth's magnetic field, freak (if orderly) micro-tornadoes and, inevitably, aliens. The banal truth: locals (and copycats around Britain) were creating the circles with wooden planks as a drinking-night prank. "Croppies" insist it's this explanation that's a hoax.

El Dorado

To Spanish explorers, he was "El Dorado," the gilded one: a god-king who founded a sacred city that bore his name and once a year was bathed in gold and rinsed in a sacred lake. In 1595 Sir Walter Raleigh led a six-week expedition 400 miles up the Orinoco River, searching for the fabled city. But El Dorado, he wrote, was always "further on"—and it still is.

The Fountain of Youth

The year was 1512, and the governor of Puerto Rico, Juan Ponce de León, now in his 50s, may have longed for his salad days, when, with Columbus, he discovered new lands for Spain. What better quest than to find the life-giving spring that Indian slaves called the Fountain of Youth? Toward the end of his decade-long search, Ponce de León discovered Florida—which, 500 years later, remains, if not a Fountain of Youth, a Magnet for the Aging.

SUNSET? A right whale dives under the waves off Peninsular Valdez along the coast of Patagonia

Photograph by Iaian Kerr — Ocean Alliance

Discoveries in
Life

" We coexist on this planet with whales, who have a brain that's not only much larger but more complex and far older than ours. And we don't have a clue what it's used for. What are the thoughts that go through such a brain? We spend millions upon tens of millions of dollars to probe the surface of Mars to see if we can find a single-cell organism or a microbe and right here on Earth is this great brain, this great intelligent form of life that might as well be extraterrestrial for all we really know of it. "

—Roger Payne

Ancient Creatures In a Hidden World

From his first day in Vu Quang, a reserve that lies on the mountainous divide separating Vietnam from Laos, biologist John MacKinnon realized he had entered an extraordinary, almost magical domain. Working out of a small army base that in earlier years had housed North Vietnamese troops, MacKinnon and a team of Vietnamese researchers set out in May 1992 on an expedition sponsored by the World Wildlife Fund. Their mission: to survey the animals in a mysterious area of moist, dense forest largely unexplored by scientists.

Returning from his first hike in that forest, MacKinnon encountered zoologist Do Tuoc, who had spent the day talking with hunters in the nearby village of Kim Quang about wild goats in the region. MacKinnon felt a flash of excitement when Do Tuoc mentioned coming across skulls with long, curved horns mounted proudly on posts in hunters' houses. "You'd better show me," said MacKinnon, for he knew of no goats of that description in Southeast Asia.

It took him just a moment with the skulls to realize that he was looking at an animal unknown to science. Subsequent analysis of the specimens' DNA showed that the 220-lb. animal, variously called the saola, Vu Quang ox and the pseudoryx, was not just a new species but a new genus, probably separated from its closest cattle-like relative for the past 5 million to 10 million years. Its narrow, two-toed hooves, concave on the bottom, provide a good grip on the local terrain.

Finding an undiscovered genus of a large land mammal was a stunning event in itself —only three other new genuses were documented in the 20th century. But MacKinnon's beast was just the first of the wonders to emerge from Vu Quang and adjoining forests in Vietnam and Laos. In the '90s scientists also found two other as-yet-undiscovered mammals in the region, the tiny leaf deer and the long-antlered muntjac.

William Robichaud—Wildlife Conservation Society (3)

NOT UNICORNS, BUT CLOSE The adult saola at right lived only 18 days after it was placed in a menagerie in a Laos. The animal's horns resemble the long uprights of local spinning wheels, inset above, also called saolas

ALAN RABINOWITZ

NEW FACES IN THE BESTIARY

In addition to the saola pictured at the far left, scientists exploring Southeast Asia discovered two other mammals unknown to science in the '90s. At top right is the tiny leaf deer, discovered in northern Myanmar in 1997. It is the second smallest of the world's deer and is the most primitive.

At bottom right is the large-antlered muntjac, which was first located within a menagerie owned by a Laotian military group. It is 50% larger than the common muntjac.

The braincases of both animals are small in proportion to their size, suggesting that the species have remained essentially unchanged for eons.

How could the natural riches of Vu Quang remain unknown to outsiders for so long? Part of the explanation lies in the region's steep, rugged terrain and exceptionally wet, sweltering weather; in addition, Vietnam and Laos were isolated for years by wars and trade embargoes.

COELACANTH
SECRETS

The coelacanth is a big fish: a hearty specimen can be 5 ft. long and weigh as much as 200 lbs. It lives among the caves and crevices along the flanks of volcanically active islands. At night, they forage for food down to depths as great as 475 ft. Foot-long coelacanth pups are born alive and fully formed.

Coelacanths and three classes of lung-fish are the only surviving representatives of an enormous group that thrived during the Devonian period, about 400 million years ago (below, a fossil). These so-called lobe-finned fish are those from

which land-based vertebrates may have evolved.

The number of living coelacanths is rapidly declining; there is a black market in their fluids, used in Asian folk medicine. The story of the ancient fish's discovery and lore is well told in Samantha Weinberg's *A Fish Caught in Time* (HarperCollins; 2000). An engaging website can be found at *www.dinofish.com*.

Meet the Dinofish

"Dear Dr. Smith," the letter began, "I had the most queer-looking specimen brought to notice yesterday … It was trawled off Culmna coast at about 40 fathoms … It is coated in heavy scales, almost armour like, the fins resemble limbs and are scaled …" The letter was written on Dec. 23, 1938, on stationery of the East London Museum in South Africa. The writer was young Marjorie Courtenay-Latimer, the museum's one-person staff, who had purchased the unusual specimen from the captain of a local fishing boat. The lettter was addressed to James Leonard Smith, a chemistry lecturer at Rhodes University and amateur ichthyologist.

When Smith solved the riddle of the specimen, his creel boasted one of the great biological finds of the 20th century. For the strange-looking fish was a coelacanth, a creature thought to have been extinct for 60 million years and known to modern science only through fossils. The finding of a live specimen in a fisherman's net was like stumbling upon a *Tyrannosaurus rex* on a golf course—except the coelacanth is far older.

The coelacanth was immediately branded the "missing link" among fish, for its sturdy fins presaged the limbs that would later develop into legs and arms in land animals. Smith searched for years—interrupted by World War II—for another, and the posters he placed on fishing wharves all over the southern African coast paid off in 1952, when a second coelacanth was found off the Comoros Islands, near Mozambique in the Indian Ocean. More were found in the following years, but the range of the exotic swimmer was thought to be tightly restricted to southern Africa. Then, in 1997, American marine biologist Mark Erdmann and his new bride, Arnaz Mehta, spotted the unmistakable form of a coelacanth in a fish cart in a market on the Indonesian island of Sulawesi—6,200 miles from the Comoros Islands. After a search that mirrored the quest of Dr. Smith 50 years before, Erdmann found a number of other specimens, and a second habitat of this living fossil, which Smith had dubbed "Old Four Legs."

Top photograph by Norbert Wu; bottom by Al Giddings

"OLD FOUR LEGS" A fish whose fins seemed as sturdy as legs "walked" right out of the past in 1938. Above, a complete specimen; below, the skeletal structure is revealed

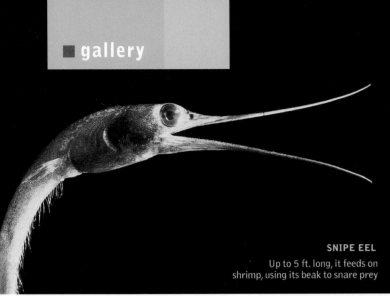

SNIPE EEL
Up to 5 ft. long, it feeds on
shrimp, using its beak to snare prey

BRUCE H. ROBINSON

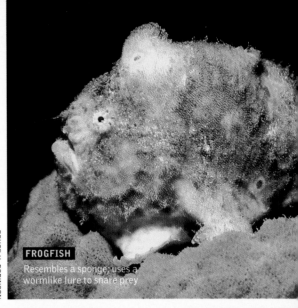

FROGFISH
Resembles a sponge; uses a
wormlike lure to snare prey

NORBERT WU

FANGTOOTH
Scary, huh? The good
news: he's only 10 in. long

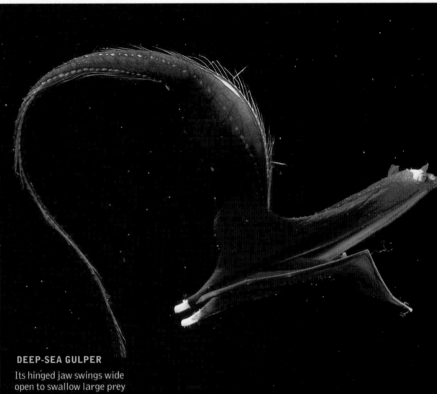

DEEP-SEA GULPER
Its hinged jaw swings wide
open to swallow large prey

NORBERT WU

Mysteries of the Deep

Deep beneath the sea, exotic life-forms adapt to a demanding environment we have only begun to explore. Here are fish that generate their own light and electricity and grow their own lures

Think of it as the earth's last frontier: the ocean world covers nearly three-quarters of the planet's surface. The sea's intricate food webs support more life by weight and a greater diversity of animals than any other ecosystem. Interest in this unexplored world began to build in the 1970s, when scientists found black clouds of superheated, mineral-rich water spewing from chimney-like mounds on the floor of the Pacific Ocean, vents between tectonic plates. Even more astonishing, scientists found that some of these submerged geysers are bursting with life. On a dive off

the Galápagos in 1977, researchers found the water around a vent teeming with bacteria and surrounded for dozens of feet in all directions with peculiar, 8-in.-long tube-shaped worms (top right, above), clams the size of dinner plates, mussels and at least one specimen of a strange pink-skinned, blue-eyed fish. On these pages is a bestiary of unusual fish that live in the deep ocean. They have evolved bizarre traits to survive, including flat, crushproof forms; glow-in-the-dark skin; translucent bodies; and built-in fishing lures. Much about their life and habitat remains unknown.

ANGLER FISH AND PREY

This "fishing" fish has a built-in
pole with a bioluminescent lure

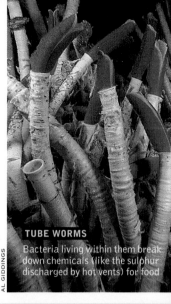

TUBE WORMS

Bacteria living within them break
down chemicals (like the sulphur
discharged by hot vents) for food

MUSHROOM SOFT CORAL

Watch out—the "palm trees"
are feeding tentacles

HATCHET FISH

It sports bioluminescent
plates along its sides

PADDLEFISH

The long paddle in front
senses electrical fields

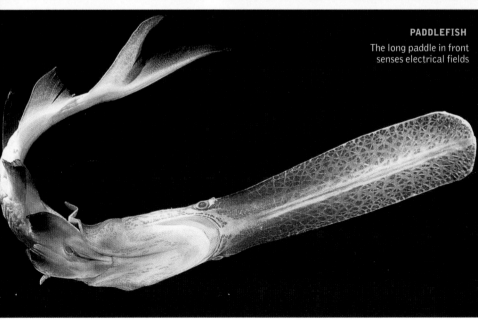

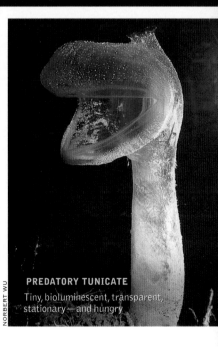

PREDATORY TUNICATE

Tiny, bioluminescent, transparent,
stationary—and hungry

Seeker of Cetacean Serenades

What was the source of the eerie, beautiful sounds that resonated at low frequencies beneath the waves? Roger Payne found the surprising answer

THE U. S. NAVY WAS WORRIED. IT WAS THE mid-1960s, and new acoustic technology had detected for the first time a low-frequency rumble permeating the world's oceans. The signal, picked up mostly at night, originated from different points around the globe and followed a tightly structured, nonrandom pattern that could come only from a highly sophisticated, intelligent source. The answer seemed clear: the Russians had developed a way of using echolocation to track American submarines.

In fact, this grave threat to national security turned out to be coming not from the Soviets but from the cetaceans—whales. Specifically, from the humpback whale. And the noises that made the Navy nervous were, of all things, love songs. "Nobody could imagine that this noise was coming from an animal," recalls Roger Payne, who in 1967 developed (along with colleague Scott Mc-Vay) the surprising theory that whales sang.

Payne, a neurophysiologist who trained at Harvard, Cornell and Tufts, spent the early part of his career studying the ways in which bats, owls and moths use echolocation to keep their bearings and locate food. But when a colleague gave him a tape of what turned out to be humpback whale songs, "I played it over and over again for a year," says Payne. Finally, with Mc-Vay, he noticed that "it was repetitive and rhythmic, and it used metered rhyme the way human music does ... it was composed according to a series of rules that are very similar to the rules human beings use when they write music." Huh? "For example," explains Payne, who is an accomplished amateur cellist, "the ratio between the atonal, percussive sounds that provide beat and the harmonic sounds that provide melody is consistent in all whale songs, and is almost exactly the same as the corresponding ratio used in our music. We also have a musical structure called the 'ABA' form, in which the composer begins with a theme, varies it into a distinct second theme, then returns to the first theme and ends the piece where it began. Whales do exactly the same thing." Strangely enough, they also modify their breathing while singing: To avoid interrupting the beat in songs that can run 30 minutes or longer, whales sneak breaths between musical phrases, much as people do.

In whales (as so often in humans), the urge to sing appears to be driven by the search for a mate. But whale songs—which have been documented only in males and primarily during mating season—still hold several mysteries. "We don't understand how they make these sounds," Payne says (one tantalizing hint: even when they breathe in mid-song, the release of air doesn't contribute to the sound). Moreover, says Payne, "we're not sure what they're using their brains, which are the largest in all of nature, for."

The search for answers has led Payne to some intriguing speculations on the atavistic appeal that music seems to hold for humans and whales alike. "I'm convinced that music is older than our species," he explains, "that it speaks to something in us that is universal—not just for humans but for all mammals and perhaps for all intelligent living things."

Payne, 65 in 2001, has a new undertaking, the Odyssey Project. "We're spending five years studying the slow accumulation of toxins in the world's oceans," he explains, "and monitoring the effect this pollution has on whales and other higher mammals." Payne hopes that this study will help save not only whales, but also a mammalian cousin to the whale (albeit with a slightly smaller brain)—*Homo sapiens*.

Close Encounters

The four-year-old girl couldn't get a straight answer from any of the grownups. It was 1938, and Jane Goodall was visiting her grandparents' farm in the English countryside, where she was given the task of collecting eggs from the henhouse. But nobody would explain how the eggs got there. So Goodall burrowed into the straw on the henhouse floor and quietly waited for four hours until a hen deposited an egg in its nest.

Fast-forward to 1956: Goodall, now in her early 20s, shows up at the Olduvai Gorge excavation of paleontologist Louis Leakey and announces that she has come to Africa to study chimps. In short order, Leakey puts her in charge of a project to identify similarities between human and primate behavior, the better to advance his thesis that chimps and people share a common ancestor (in addition to sharing 98% of their DNA).

After spending frustrating weeks alongside the chimps at her Tanzania camp, Goodall hits pay dirt: she observes an adult male chimp she had named David Graybeard strip the leaves off a twig and push it into a termite mound to gather up insects, which he then sucks off. The idea of animals making tools was something that both scientists and theologians of the day said was impossible. Ditto the emotional and social behavior she later saw chimps display—tickling, kissing and hugging—as well as less enchanting (and all too human) habits: sadistic murder, cannibalism and wholesale genocide.

After more than four decades in the bush, Goodall today spends much of her time trying to sound the alarm about a different kind of genocide: in the early years of this century, there were almost 2 million chimpanzees living in 25 African countries; now there are fewer than 100,000 left in just four countries. "To save what you love," she says, "you have to leave what you love sometimes." But only after observing it closely.

Michio Hoshino—Minden Pictures

CHEWING IT OVER Jane Goodall enjoys a quiet moment with a favorite chimp at her camp in Gombe, Tanzania

Like Jane Goodall, American Dian Fossey was a protégé of Louis Leakey's. The two met memorably: visiting Leakey's Olduvai Gorge dig site, Fossey ran down a hill to see a newly unearthed find, but fell and broke her ankle. "The sudden pain," she later said, "induced me to vomit unceremoniously all over the treasured fossil."

Nonetheless, three years later, Leakey invited Fossey to take over a project studying great apes. In the course of 15 years of close observation, she discovered that gorillas are considerably less violent than previously imagined (they avoid physical confrontation, unless their young are endangered) and far more social: Fossey was able to demonstrate that they live in stable social groups of up to 20 (led by an older male, called a "silverback") for long periods of time. She also found that gorillas are vegetarians and that —sadly, we will never know the rest of what Fossey might have discovered. In 1985 she was murdered in her Rwandan camp, at age 53. Her killing has never been solved.

NEW TOOL FOR A NEW WORLD

From 1573—when Sir Frances Drake climbed a tree in the Serranía del Darién in Panama and got his first look at the Pacific Ocean—until the 1970s, exploration of the rain forest canopy was slow. A few researchers began to revolutionize canopy science in 1974 by using rope-climbing techniques, but only the pluckiest —like Don Perry, perched 100 ft. above the ground in a ceiba tree, above—get close to the creatures and plant life they're after. Others have tried ladders, towers, suspension bridges, walkways and rubber rafts dropped into tree-tops from dirigibles.

Finally, in 1990, came the canopy crane, an ordinary construction crane fitted with a gondola that can be run along the horizontal arm and then lowered by pulley into the treetops, providing a view of living things while having a mini-mal impact on fragile ecosystems. The first was set up on Pana-ma's Pacific Coast; now others are in service in Costa Rica, Venezuela and Washington State.

 **environment**

Brave New World

Welcome to a mysterious and still largely unexplored realm, the rain forest canopy. This treetop network of leaves, vines and branches forms a world within a world. In these huge green expanses far above the forest floor, sky gardens of orchids and philodendrons bloom without benefit of soil. Birds avoid the rain by hanging upside down beneath broad leaves. Green iguanas climb aloft to bask nearer the tropical sun. Sloths dangle motionless for hours, avoiding eagles and other predators. And everywhere are the buzz and flutter of insects, which inhabit the rain forest in incredible variety and numbers. And don't forget the buzz and flutter of scientists racing to plumb the biological secrets of this world before it disappears.

The canopy is physically and biologically the most active part of any rain forest. With its complex system of interdependence among animals, plants and insects, it is also the least understood. Although it may appear to be a conservationist's cliché that rain forests must be preserved because a cure for cancer may be found there, experts believe that these tropical regions still harbor substances that can help alleviate many forms of human suffering. One of the oldest and most notable medical success stories, from the tropical forests of Madagascar, is the rosy periwinkle. The delicate pink flower is the source of two of the most powerful anticancer drugs. Plants "bioprospected" in Cameroon, the Central African Republic, Tanzania, Gabon and elsewhere in Africa have yielded compounds that have tested well against HIV and malaria. The Amazon may yield a treatment for paralysis based on the venom of a spider, and an anti-coagulant derived from the saliva of a bat.

More such discoveries are still in store—if the rain forests survive long enough for researchers to study their bountiful flora and fauna. Some 6% of the earth's land area is rain forest, and more than 1% of that—some 42 million acres—is being lost each year through logging, mining and agriculture.

Photograph by Mark Moffett—Minden Pictures

MECHANICAL STILTS A canopy crane high above the rain forest floor in Costa Rica allows researchers to explore an area only recently accessible to man

PHANTASMAL POISON-DART FROG
An extract from the skin of this tiny Ecuadorian tree frog is a more potent pain killer than morphine (at least in rats). Stay tuned!

Potions from Poisons

A host of new drugs emerges from an unexpected source: the natural world's most toxic substances

If you're prospecting for new drugs in nature, there's no better place to start than the business end of a good poisonous plant or animal. Modern medicine is filled with drugs derived from deadly poisons, from the muscle relaxant curare (taken from South American vines that are used to poison arrow tips) to the anticoagulant Aggrastat (based on the venom of the saw-scaled viper).

Animal venoms make particularly good sources of potential drugs because they are designed to kill or immobilize prey. Many contain dozens or even hundreds of potent, fast-acting toxins that home in on the muscles and nervous system. The molecules also tend to be small, which means they can easily slip across the blood-brain barrier, the network of tiny vessels in the brain that blocks larger compounds Poisonous snakes, spiders, scorpions and frogs have so far attracted the most scrutiny, but insects and marine creatures are also rich sources of potent compounds. Here's a taste of what's going on:

GIANT ISRAELI SCORPION
Chlorotoxin, a substance in the venom of the 5-in.-long species known as the "death stalker," may offer hope for the 25,000 Americans each year who have glioma, an incurable form of brain cancer

TERCIOPELO SNAKE
This Central American pit viper's venom contains a compound that fights bacterial infection

THAILAND COBRA
Immunokine, a drug derived from its venom, is used to combat multiple sclerosis

JOE MCDONALD—ANIMALS ANIMALS

***PHYLOBATIS TERRIBILIS* FROG**
Its poison, used by Central American tribes to disable prey, is yielding new antibiotics

SOUTHERN COPPERHEAD
Its venom is a powerful clot buster and may retard the growth of cancerous tumors

JOYCE & FRANK BUREK—ANIMALS ANIMALS

CONE SNAIL
This tropical swimmer stabs its prey with harpoons loaded with poison; the venom may help treat schizophrenia, stroke and Alzheimer's

MARTHA SWOPE—TIMEPIX

Chronicler of Great Divides

Whether tracking sexual development in the South Pacific or taking sides in the sexual revolution in the U.S., Margaret Mead made anthropology matter

A REPORTER TURNING UP AT ONE OF HER lectures noted that the speaker had somehow managed to discuss stones, museums, stuffed birds, dinosaurs, whales, cave painting, Cro-Magnon man, the possibility of life in outer space, the oneness of the human species, education, pollution, evolution, growing up in New Guinea, relations between the sexes, common names and the fragmentation of communities.

The list was typical: Margaret Mead got around. Her studies—and the two dozen books that resulted from them—revolutionized her chosen field of anthropology. Long before her colleagues recognized the validity of her approach, she studied the biological, psychological and sociological forces that shaped personality in primitive cultures, then used her findings to explain how individuals learn adult roles in modern societies. Her application of this approach to other areas and her willingness to speak out on almost any subject made her ideas—and her smallish but somehow imposing figure topped by its Buster Brown hairdo—famous around the world. By the time of her death at 76, Margaret Mead had become the grandmother of the global village, an all-wise matriarch whose often provocatively put, commonsense opinions were sought by millions.

Both observation and involvement came naturally to Mead, who was born in 1901 to parents who quite literally raised her to be a social scientist. She was only eight when she was assigned to observe and record her younger sisters' speech pattern. After training at New York's Barnard College and Columbia University, she sailed for Samoa in 1925 and spent nine months observing the adolescent girls of three small coastal villages in the Manua Islands. The result

of her study was published three years later as *Coming of Age in Samoa.*

The book, which described the easygoing, neurosis-free island way of life and suggested that the Western attitude toward sex could be relaxed without endorsing promiscuity, was an instant success. Many of the young researcher's colleagues condemned her way of reaching conclusions from observed evidence, which Mead called "disciplined subjectivity." But students snapped it up, partly because its ideas interested them, often because, as the author briskly explained, "I wrote it in English."

Mead wrote her other books in the same easily understood idiom. But anthropology alone could not satisfy her. A fluent speaker who rarely needed notes, she also carried a heavy teaching schedule. She established a Hall of the Peoples of the Pacific at the American Museum of Natural History, where she was curator of ethnology. She brought her keen mind and anthropological insights to bear on her own sociey, and spoke out—thoughtfully, candidly and, some said, too frequently—on social problems that many of her colleagues preferred to avoid.

When Mead appeared on TV talk shows to endorse women's liberation or other causes, envious colleagues said she was "overexposed"; conservative academicians called Mead, who served on more committees than anyone could remember, an "international busybody."

Mead carried on. She had survived malaria, three marriages and years of native foods, but in 1978 she succumbed to cancer. Characteristically, she turned her trained observer's eye upon herself and recorded her own progress of aging. Her attention was appropriate. Of all the people she studied, few were as interesting as Margaret Mead herself.

Margaret Mead in Samoa, 1926

LAST DAYS OF
THE KORUBO

There are few, if any, truly "lost" tribes anymore. Among the most recent to be "contacted" is the Korubo, who live along the Itacoaí River in Amazonas state in western Brazil.

The existence of the Korubo has been known since 1973, but it was only in October 1996 that agents of the Brazilian government's National Indian Foundation (FUNAI) met with perhaps 20 of an estimated 200 to 300 of them. Previous attempts to contact the seminomadic hunter-gatherers resulted in clashes in which nine FUNAI agents and an untold number of Korubo were killed.

As loggers moved into the area, the Korubo realized they needed allies. FUNAI agents left machetes and aluminum cooking pots on the ground, then withdrew. Deciding to trust the visitors, several men emerged from a corn field to collect the booty and to offer ears of corn in return. Such gifts may foster dependence, but they are less threatening than the alternative technology the Korubo face: loggers' guns.

Better Lost Than Found?

Victorian explorers called them "lost tribes"; contemporary scientists call them "indigenous peoples." Yet all too soon, these people may indeed be lost tribes again—forever. Expelled from the planet by civilization's sprawl, they will take with them knowledge we are only beginning to appreciate. Stored in the memories of elders, healers, midwives, farmers, fishermen and hunters in the estimated 15,000 such cultures remaining on earth is an enormous trove of wisdom. Over the ages, indigenous peoples have developed innumerable technologies and arts. They have devised ways to farm deserts without irrigation; they have learned how to navigate vast distances in the Pacific using their knowledge of currents and the feel of intermittent waves that bounce off distant islands; they profit from the medicinal properties of little-known plants. Much of this expertise and wisdom has already disappeared, and if neglected, most of the remainder could be gone within the next generation.

The pace of change is startling. The semi-nomadic hunters and gatherers of Borneo's Penan tribe dwindled from 10,000 at the beginning of the 1980s to fewer than 500 by 1990. Today, only some 260 Penan are estimated to survive on the island. Since 1900, 90 of Brazil's 270 Indian tribes have completely disappeared, and others, like the Korubo people at left, are close to extinction.

The most intractable aspect of the crisis is that it is largely voluntary. Entranced by images of the wealth and power of the First World, the young are turning away from their elders and their ways, breaking an ancient but fragile chain of oral traditions. Penan villagers know that their elders used to watch for the appearance of a certain butterfly, which always seemed to herald the arrival of a herd of boar and the promise of good hunting. These days, most of the Penan cannot recall which butterfly to look for.

Photographs by Erling Söderstrom (2)

ENDANGERED A group of Korubo women and children in Brazil; while Korubo women wear no clothing, the men wear penis shields and headbands. Top left, a Korubo village

Regal Retreat

The annual migration of the resplendent monarch butterflies from their summer homes in the U.S. and Canada to their wintering spots in Central America is one of the great sagas of nature: these steadfast insects brave wind, rain, sleet, snow (but not dark of night, for they are day-flyers only), enduring a journey of 7,000 miles or more to reach their winter havens.

For decades, the monarch migration was also one of the great mysteries of nature, for no one knew the location of the butterflies' winter retreat. When Canadian zoologist Fred A. Urquhart found the answer, his discovery climaxed four decades devoted to studying the brilliant flyers. Urquhart first began testing ways to tag butterflies in 1937; after many years of trial and error, he finally discovered an adhesive strong enough to bond to a monarch's wing yet light enough not to damage it. In the early 1950s he began seeking volunteers to help him tag the monarchs and track their migration; his Insect Migration Association eventually numbered in the thousands.

In 1973 Urquhart received a letter from Kenneth Brugger in Mexico City, who volunteered to to crisscross the Mexican countryside in search of the monarch haven. In 1975 Brugger found his first monarch winter colony in a 20-acre patch of fir trees high in the Sierra Madres; Urquhart joined him there the next year. The insects were in a state of semidormancy, sleeping deeply and covering fir trees in the tens of thousands. Other wintering places of the monarchs have since been found, each at a height— about 9,000 ft.—where the cool winter air sends these poikilotherms (animals that adjust their body temperature to the surrounding air) into a long winter's nap.

Photographs by Bianca Lavies—National Geographic Society (2)

WINTERIZED The monarchs begin to stir when the sun warms them; on his first visit, Urquhart found a butterfly bearing one of his tags that had migrated from Minnesota

SEASONS OF THE MONARCH

A monarch sports one of Fred Urquhart's tags, above. In his long years of research on the migrating insects, Urquhart learned that the monarch population in Canada includes both migrant and nonmigrant insects. Urquhart concluded that most of the migrating monarchs hatch in late summer, as the shorter days of fall set in. Females born at this time do not become sexually mature until they have wintered in the warm climes. When they awaken from their semidormant state, the females, now sexually mature, begin to fly north, as they feel the urge to mate. Almost all males of the species die on the return trip north.

Females deposit monarch eggs only on milkweed plants. A monarch larva sheds its skin five times as it grows: the process of maturation from egg to larva to caterpillar to pupa to adult takes about five weeks.

A grand question remains: How do the butterflies born in Canada "know" the direction to the winter haven in Mexico? Urquhart and others suspect the insects are sensitive to the area's high magnetic field.

MOUNTAIN GORILLA
Congo, Rwanda, Uganda
Number: About 350

Death Row

The extinction crisis hits humanity's family tree, as a bevy of primates is now deeply endangered

As far as we know, no primate became extinct during the 20th century. That's an impressive record, since the world is currently losing about 100 species a day. But luck may soon run out for the animal order that includes humans. As their habitats are destroyed by human population growth, dozens of our closest relatives—from the gorillas in the mists of East Africa to the wise-looking orangutans of Sumatra—are on the brink of oblivion. When the new century began, Conservation International, a private group based in Washington, and the Primate Specialist Group of the Species Survival Commission of IUCN/The World Conservation Union, an international environmental alliance, compiled a list of the 25 most endangered primates. Eight of them are featured here; some of the others are so rare that no pictures of them exist. Does one of them offer clues as to how humans evolved? Does another harbor natural antibodies that fight cancer or AIDS? We don't know—and once they're gone, we'll never know .

FRANS LANTING—MINDEN PICTURES

SUMATRAN ORANGUTAN
Indonesia
Number: Fewer than 5,000

GOLDEN BAMBOO LEMUR
Madagascar
Number: About 1,000

GOLDEN LION TAMARIN
Brazil
Number: 800 in
the wild; 500 captive

BUFF-HEADED CAPUCHIN
Brazil
Number: Fewer
than 5,000

DUOC LANGUR
Vietnam
Number: Fewer
than 1,000

DRILL
West Africa
Number: Unknown

GOLDEN-CROWNED SIFAKA
Madagascar
Number: Fewer than 5,000

It's Alive! — Or Is It?

We don't believe in unicorns and dragons anymore —
but don't tell the tourists there's no monster in Loch Ness

Biosphere II

Remember Biosphere II, the three-acre sealed Arizona complex designed to house half a dozen "bionauts," plus hundreds of species of plants and animals in a state of biological equilibrium, from 1991 to 1993? This "natural" environment was sustained by $30 million worth of computers, filters, pumps and recycling equipment. But the oxygen systems failed, the crops died and the people resorted to smuggling in supplies. Perhaps it was just ahead of its time: it would have made a great reality-TV show.

Loch Ness Monster

The legends date to A.D. 565, when St. Columba was said to have rescued a farmer from a monster's grip near a Scottish lake, Loch Ness. In 1868 a local newspaper reported sightings of a huge fish. In 1934 a grainy image of the "Loch Ness Monster," right, was snapped by a local doctor. Never mind that in 1994, that shot was finally revealed to have been a prank. As tourism surpasses sheep ranching as Scotland's No. 1 industry, what nobody can deny is that the millions of dollars "Nessie" earns each year are for real.

HULTON/ARCHIVE

RALPH WHITE—CORBIS

Midwife Toad

Darwinian orthodoxy says adaptation happens randomly and very slowly. But in the late '60s, Austrian biologist Paul Kammerer published research on the midwife toad suggesting that animals could develop a selective advantage in a single generation and pass it on to their offspring. But his data were faked: shortly after being exposed, he committed suicide.

Bigfoot

He is said to love fresh fruit and emit a strong odor—but then so might you if you weighed 500 lbs. and stood almost 10 ft. tall. Believers say Bigfoot ("sasquatch" to Native Americans) is a primitive hominid that somehow found its way to the Pacific Northwest tens of thousands of years after its brethren had become instinct. Skeptics contend pictures like the one above are hoaxes.

Trofim Lysenko

The Soviet agronomist claimed he could make spring wheat into winter wheat, by "training" it to pass along acquired characteristics, a notion that appealed to Joseph Stalin. But Lysenko failed, and in failing, he doomed a generation of Soviet scientists to missing out on the genetics revolution.

Abominable Snowman

In 1892 a British diplomat in Nepal wrote home about natives on his staff who were frightened by *bun manosh*— wild men. Since then, the "yeti," or abominable snowman, has been sighted across central Asia. In 1951 Everest climber Sir Edmund Hillary, inset, found fresh footprints near the Tibet-Nepal border that were longer and wider than a climbing boot. But no one has ever brought back a yeti, live or dead. One suspect: the "chemo," a rare, 7-ft.-tall Tibetan bear whose footprints may resemble a primate's.

index